INSURING QUALITY

How to Improve Quality, Compliance, Customer Service, and Ethics in the Insurance Industry

Hedy Abromovitz
Les Abromovitz

St. Lucie Press
Boca Raton, Florida

Library of Congress Cataloging-in-Publication Data

Catalog information may be obtained from the Library of Congress

St. Lucie Press is an imprint of CRC Press LLC

International Standard Book Number 1-57444-150-7
Printed in the United States of America 1 2 3 4 5 6 7 8 9 0
Printed on acid-free paper

Table of Contents

PREFACE

At the very least, the insurance industry has an image problem. A recent article in *The New York Times* called it an industry the public hates. The article gave examples of insurance company employees who were embarrassed to admit what they did for a living.

In books, magazines, and movies, insurance companies and their agents are maligned. In his best-seller *The Rainmaker,* John Grisham's plot revolved around an evil insurance company and its employees.

Woody Allen in his movie *Love and Death* said, "There are things worse than death. You could have to spend an evening with an insurance salesman."

There's even a greeting card that gets into the act. The card relates what looks to be a story problem like all of us encountered in high school. It goes something like this:

> A man sitting next to an insurance salesman leaves New York on a high-speed train. Another man also sits next to an insurance salesman but leaves by plane. The question is: Which man will jump out the window first?

Those jokes focus on the fact that insurance is boring to most people. Unfortunately, the industry has a bigger image problem than just being boring. Recent investigations into the life insurance sales practices of Metropolitan Life, Prudential, and others have caused many to question whether members of the industry are trying to cheat them.

For an industry built on trust, that's a problem. It's imperative that consumers trust the person who's selling them insurance.

In another era, people weren't as skeptical about the insurance industry. On "Leave It to Beaver," Ward Cleaver was a great father who worked for an insurance company. No one could ever imagine Beaver's dad acting unethically in his dealings with the public.

When he wasn't at work at the insurance company, Ward Cleaver helped Beaver out of jams and taught him how to be an ethical human being. All in all, Ward was pretty neat for a grown-up, as Beaver and Wally liked to say.

Unfortunately, the real world isn't Beaver's home town of Mayfield, Ohio, or even Mayberry. There are too many Eddie Haskells running around trying to scam the rest of us. Many members of the public perceive that it's Eddie Haskell working for the insurance company, not Ward Cleaver.

"I've been paying premiums for 30 years," the customer says with only a year or two of exaggeration. "And now you're not paying me? This policy isn't worth the paper it's written on."

That's the customer's perception, and you'll have difficulty persuading her that she's incorrect. She's paid out thousands of dollars in premiums and is getting no visible benefit.

Insurance companies have to sell more than just a piece of paper. It's image, strength, security, protection, and peace-of- mind that they're selling. When they purchase insurance, consumers are buying an intangible product.

The American Council of Life Insurance (ACLI) has taken aggressive action to remedy the dire situation. It has developed six Principles of Ethical Market Conduct and an accompanying Code of Life Insurance Ethical Market Conduct. These new standards of ethical market conduct are designed to improve the industry's image and eliminate the market conduct problems that have plagued insurers.

Some of the best advice on improving quality and customer service is contained in the new Principles and Code from the ACLI. Many of the standards contained therein would work well with other lines of insurance.

The new ethical market conduct standards from the ACLI utilize guidelines that are similar to the Malcolm Baldrige National Quality Award criteria. In fact, Baldrige assessors may be used in the evaluation process.

It's not just the ACLI that recognizes there's a problem. The Life and Health Insurance Foundation for Education (LIFE) was formed in 1994 by seven life and health insurance groups. LIFE's purpose is to "educate the public about the invaluable role of life and health insurance agents, companies, products and services."

It's not just life and health insurance products that are invaluable. Few people would dispute the value of automobile or homeowners insurance.

So why does insurance get a bad rap? A partial explanation is that insurance isn't associated with the good times in life. You usually need it when something bad happens.

But quality has a lot to do with it too. And by improving quality, as well as service, the industry can improve its image and show the public that it is offering a truly invaluable product.

Throughout this book, we'll be examining the connection between quality, customer service, ethics, and compliance. They go hand in hand and sometimes cannot be separated. The insurance industry must make improvements in those areas.

You probably know how to use the standard quality improvement tools. You can use those tools to implement improvements in the problem areas identified.

The ultimate answer for your company or organization isn't necessarily spelled out in this book. Each short chapter is a starting point, which provides an overview of a problem, as well as suggestions for improvement.

Ideally, you'll utilize the material in training sessions as a springboard for discussion. You'll see quickly if the same or similar problems exist in your organization. Employees at your company can come up with improvements to remedy the problems discussed in the book, as well as their own.

This is not a book for organizations that like spending hours, if not days, in meetings and training. It's not for those companies that like complicated diagrams. You won't find any fables or allegories. Every problem in this book is based upon a true customer-service nightmare drawn from our version of "The X-Files."

You won't deal with hypothetical situations that only happen in the classroom of your college philosophy professor. You won't be required to make deep philosophical choices, like who you save from an earthquake—your spouse or the scientist who has found a cure for cancer.

Insuring Quality will help your organization solve real-life problems. It's for the people in the trenches, not for those who want to pontificate about quality and service from an ivory tower.

Insuring Quality offers practical solutions to tough customer-service problems using quality improvement methods. It can help resolve conflict when no win/win options are available.

Quality and customer service can be improved. And you've picked up a book that can help. It can help the industry pick up the pieces and start doing things right.

No, the industry is never going to be beloved by the public, but it will be respected and trusted. *Insuring Quality* can help insurers rethink their positions on quality, ethics, compliance, and customer service. And by doing so, it can lead to the public rethinking its opinion about insurers.

ACKNOWLEDGMENTS

We've always believed that quality and service would improve if every insurer treated customers like they'd want members of their own family treated. For over two decades, our insurer, USAA, has treated us that way. It is an insurance company that should be an inspiration to others. We've been blessed with loving families who've always inspired us to do the best we can. We're extremely fortunate to have parents who are always in our corner. We've also been inspired by the Engleman family, who fought the system on behalf of their daughter, Haley, and helped bring about expanded coverage for Enteral Formulae.

About the Authors

Hedy Gruenebaum Abromovitz, a chemist by training, is Vice President of Quality for a company that manufactures audio and video surveillance equipment for law enforcement. She has spent over fourteen years in Total Quality and Quality Assurance functions in a variety of industries. She has written about quality and customer-service topics for technical journals and many other publications.

Les Abromovitz is an attorney and author of *Family Insurance Handbook,* which was published in hard cover and paperback by McGraw-Hill. He has been the editor of several insurance books and analyzes insurance cases for a legal newsletter. Les is the author of a legal humor book, as well as several other books.

Les was an insurance regulator for two years and worked for a major insurance carrier for almost fifteen years. He is the associate producer of a national series on public television dealing with insurance, legal, and personal finance subjects.

Hedy and Les are co-authors of *Bringing TQM on the QT to Your Organization: How to Implement Total Quality Management Without Shoving It Down Your Employees' Throats,* which was published by SPC Press. The couple split their time between Pittsburgh, Pennsylvania, and Boca Raton, Florida.

CHAPTER 1

The Relationship Between Quality and Image

The insurance industry really cares about quality and customer service. Some companies even go so far as to build an ad campaign around it.

They want the public to know how much they care. The ad will state that "Quality is our policy" or that the company has "A Commitment to Quality Service." Another insurance company brags that it has been "Setting Standards in Excellence Since 1925."

Companies not only talk about quality; they sing about the service you get from them. Sing it with us now: "Like a good neighbor, State Farm is there."

Or hum the tune of "You're in good hands with Allstate." Harmonize with us as we sing, "Coming through for you. That's what CG people do."

All of them are good companies with catchy songs. But customers don't always sing a happy tune after dealing with their own insurance company or agent. If customers were consistently singing the praises of their insurance company, the industry's image wouldn't be what it is now.

The Wall Street Journal reported on November 7, 1995 that insurance companies suffered the largest drop in customer satisfaction, according to the American Customer Satisfaction Index. The survey was conducted by the American Society for Quality Control, a nonprofit organization based in Milwaukee, Wisconsin.

The customer satisfaction rating of insurance companies dropped 6.2 percent. Some of the decline in customer satisfaction might be attributed to nonrenewals and rate hikes in Florida as a result of hurricanes. Or maybe, just maybe, quality and customer service in the insurance industry have declined significantly despite ongoing efforts to make improvements.

While it's just one survey, almost everyone would agree that there's room for improvement in the insurance industry. Even if the American Customer Satisfaction Index showed an increase in customer satisfaction, there would still be room for improvement.

A lot of room.

Although you can't please every customer, it's not natural disasters that have caused people to be dissatisfied with the insurance industry. Natural disasters bring out the best in the insurance industry. Frequently, insurance company employees spend months away from home, helping people put their lives back together in areas hit by hurricanes, earthquakes, and floods. They help because it's their job and because these insurance company employees believe they're doing something worthwhile. And they truly are performing a remarkable service.

It's the day-to-day problems, not the natural disasters, that have led to customer dissatisfaction with the insurance industry. Customers are dealing with overly aggressive salespeople, busy customer-service lines, incorrect bills, untrained employees, missed appointments, slow claims response, and a host of other problems.

When customers need to call on their insurance company or agent, they're not in the best of moods. In many cases, the insurance industry is catching them on their worst day. Somebody scratched their car, or they're scratching for the money to pay the latest hike in their auto insurance premium.

The life insurers have their own image problems. It starts when customers open up their morning newspaper and turn to the business section.

The lead story is that Prudential Life Insurance Company of America has been fined $35 million for deceptive sales practices by a task force of 30 state insurance regulators. An estimated $1 billion will be paid to consumers as part of a restitution program. There are also class action lawsuits against the company.

Unfortunately, there's more bad news affecting the life insurance industry. In personal finance magazines and on television, there are negative stories about unethical agents and companies.

According to research done by the American Council of Life Insurance (ACLI), the industry's image has suffered from the adverse publicity. The research shows that the life insurance industry's image is far more negative than it once was. The ACLI research revealed the following:

- In 1994, only 45 percent of the population agreed that the life insurance industry is humane and understanding in its dealings with the public. The percentage declined from 79 percent in 1967.
- In 1994, only 24 percent of the population agreed that life insurance companies had high ethical standards. In 1968, 50 percent of the population agreed with that statement.
- In 1994, only 18 percent of the population believed "honest in dealing with their customers" to be an appropriate description of life insurers. The percentage dropped from 37 percent in 1986.

Worse yet, those statistics may not be reflective of the industry's current image. Since 1994, there have been additional exposés of unethical sales practices, as well as more litigation and actions taken by state regulators.

The public's opinion of the insurance salesperson has also declined. The ACLI research found the following:

- In 1993, 39 percent of the population agreed that life insurance agents exert no more pressure than other salespersons. In 1968, 57 percent believed that there was no more pressure from a life insurance agent than other salespersons.

- In 1994, 49 percent of those having contact with an agent felt pressured, and only 45 percent believed the agent was truly concerned about their needs.

While these statistics may not accurately reflect the public's opinion of casualty-property agents, they certainly indicate there's a problem that must be addressed. Improvements in quality and customer service will help improve these negative perceptions about the industry.

CUSTOMER EXPECTATIONS

The conduct of a few insurance industry members has altered the perception of the industry and its products. Prospects are wary of the insurance salesperson, thanks to the actions of a few.

The customer's perceptions may never be changed, no matter how ethical a particular salesperson is. The prospect will continue to view the salesperson with distrust.

It is the customers' perception of how they are treated that is extremely important. Furthermore, if the industry is perceived in a negative way, it really doesn't matter if that perception is accurate.

The climate is one of hostility toward the insurance industry. The salesperson may never have the opportunity to meet or exceed the customer's expectations. Agents may face more resistance than usual in arranging meetings with clients and prospects.

The ACLI statistics quoted above help define customers' expectations of the industry and the salesperson. Customers want:

- To be treated humanely and with understanding.
- To know that members of the industry have high ethical standards.
- To be dealt with fairly and honestly.
- To be assured the salesperson is concerned about their needs and is not pressuring them into buying.

- To have marketing materials that are clear and not misleading.

MARKET CONDUCT

Market conduct has a significant impact on the image of the insurance industry. With each report of unethical market conduct, the image of the industry deteriorates further.

Accusations of deception taint the industry as a whole, not just the individuals involved. According to a study done by Price Waterhouse in cooperation with the ACLI, deceptive marketing practices fell into three basic categories:

- Agents who disguised life insurance products as other investment vehicles.
- Agents who did not disclose that they were selling life insurance and presented themselves as financial advisors or used a sham title.
- Agents who failed to fully explain the impact of loans, early surrender, and withdrawals that can have a negative impact on the policy.

In most cases, agents did not encourage loans, early surrender, or withdrawals so that the policyholder could buy a new car or pay a child's tuition bill. The purpose was to fund a new policy on which the agent would collect a commission.

THE ECONOMICS OF THE INDUSTRY

The life insurance industry is changing significantly. In order to be more competitive, the industry has introduced products to attract investors, rather than just savers. Variable annuities, variable life, and variable universal life all offer the possibility of greater returns than more traditional life insurance products.

Products like these have attracted customers who are willing to accept more risk in exchange for a better rate of return on their investment. As risk shifts to the customer, however, the insurer's profit margin narrows.

In addition, demand for life insurance policies is flat. People who might have purchased life insurance in another era are now plowing that money into mutual funds or some other financial product. The market conduct problems of the industry may also be having an impact on demand for policies.

The distribution channels have changed too. The Supreme Court has cleared the way for banks to sell insurance. Banks can tap into their existing customer bases to obtain prospects for insurance products.

Even within the industry, the traditional insurance salesperson faces stiff competition. Customers can get life insurance quotes on the Internet. Or a customer can call 24 hours a day for a free personalized quote. The advertisement for these services states that there is no need to meet with a sales agent.

SelectQuote advertises that it's changing the way America buys life insurance. With just a phone call, customers can get quotes on term life insurance from top companies. A commercial points out what the company feels is a major selling point: "Your doorbell never rings."

Perhaps it's just a marketing pitch, but it seems that even people who sell insurance are aware of the industry's image. They believe that people dread the thought of an agent at their door.

It's not the image the industry wants nor deserves. But it will remain that way until quality and customer service improve.

Q-TIPS

The insurance industry is at a crossroads in its relationship with consumers. It can either watch its image deteriorate further or improve quality and customer service.

Chapter 2

Quality and the Media

Sure, the insurance industry has a bad image. Insurance industry executives bemoan that fact. Many of the organizations to which they belong are trying to improve that negative image.

They sit around at meetings, wringing their hands over how people hate the industry. They resent being portrayed in the media as callous and evil people. The industry provides a valuable service, but no one seems to recognize that fact.

While they're discussing the problem, a family that can use their help is in a desperate situation. Their four-year-old daughter requires a special formula to live. The formula is fed through a tube in her stomach. Without it, she will die. Her mother doesn't work because her day is spent caring for her daughters. Dad earns about $25,000 per year. They do have a better-than-average health insurance policy, however, through his employer.

Unfortunately, the family is told that the formula isn't covered by the policy, even though it's not specifically excluded and the child suffers from a rare medical condition. Nevertheless, the health insurer would pay for that same formula if the child were receiving it in a medical setting, even though that alternative would be far more expensive.

The family files a complaint with insurance regulators in their state. The company's position is upheld on the grounds that the formula is food, and food isn't covered by the policy.

A lower level insurance department employee argues that there is no specific exclusion for food in the policy, so it should be covered. His boss scoffs at the argument, saying that T-bone steaks aren't covered and that there's no exclusion in the policy for them either. Everyone, however, agrees that the formula is medically necessary.

The family gets to the point where the only option is Medicaid, which covers part of their costs. They still believe strongly, however, that their private health insurance policy should be covering the formula. Several insurance coverage experts share that opinion, and a few other carriers in the region do cover the formula that the little girl needs.

The story of the child's plight makes the newspapers and news shows. Eventually, after pressure from legislators, regulators, and the public, the health insurer decides it has an image problem. It provides a grant to the family to cover the cost of the special formula.

At that point, however, the damage is already done. The industry took another blow to its image and reinforced the opinion of many that it is heartless. The company could have made the grant money available to the family long before it was pressured to do so.

Just as the public doesn't like the insurance industry, the industry seems to dislike the media. At many carriers, the media is treated with disdain, even though the reporter is simply doing a story on insurance to educate the public. Media requests are shuffled from one department to the next. The switchboard operator has no idea who should get the call. The free publicity goes to a company that knows how to develop quality relationships with the media.

When you read about insurance in magazines and newspapers, you'll see a lot of the same names. These are people who make themselves available to reporters and respect their deadlines. They know that reporters have significant time constraints when writing a story. Ninety-nine percent of the reporters don't have an axe to grind. They're simply trying to provide an objective report on a particular subject.

Too frequently a news story will be written with no assistance from the industry or with very little input. The reporter never gets a return call from someone at the insurer, despite leaving messages on several occasions.

Customer service means taking care of all your customers, and that includes the media. It doesn't mean manipulating the truth to further a particular agenda of an organization or company.

A top official for an organization of insurance agents receives a press release for a legal humor book. The book is a tongue-in-cheek look at the litigation explosion and pokes fun at the tendency in our society to sue for any reason.

Attached to the press release is a copy of the book's cover that shows a beleaguered defendant, surrounded in the courtroom by a member of a motorcycle gang, animals, and a host of others who are suing him. In the cover artwork, all of the cartoon characters are pointing their finger at the sympathetic defendant.

The executive vice president of this insurance organization doesn't bother to call the publicist who sent the press release. He never even reads the book. Had he looked at the first page, he would have seen that the book is a spoof. Had he read even a chapter, the executive would have known that the author was discussing lawsuits involving pets, sitcom characters, inanimate objects, and people who exist only in cartoons.

Instead, the executive writes a scathing editorial about the book. He cites it as an example of another greedy lawyer promoting litigation.

Even after several readers inform him that he missed the point, the official never issues a retraction. He never apologizes or corrects the misinformation published in his organization's insurance magazine.

Perhaps the executive wanted to come out strongly against frivolous litigation, so he distorted information to prove his point.

Assuming the executive has a higher IQ than a marshmallow, the inescapable conclusion is that he purposefully misled his members to rile them up about the need for tort reform. Maybe he rationalized that the end justified the means.

While tort reform may be an important issue to some members of the insurance industry, there is no justification for making misrepresentations to bolster the need for legislation. The executive's actions were an insult to all of the customers of his organization, internal and external.

Sadly, most of us do expect magazine articles, even editorials, to be based upon some grain of truth. Whereas we know to be skeptical in reading the tabloids, we expect insurance industry publications to be accurate.

Aside from the executive's lack of journalistic ethics, there's a quality issue here. Quality is undermined when the people who put out industry publications have an agenda.

The insurance industry has every right to convey its position on issues to the public. For example, the industry seems to have finally made headway in convincing people that insurance fraud hurts everyone. Many news stories are finally focusing on the people who cheat the system, not on attacking insurance companies for failing to pay those claims.

On the other hand, Charles Givens, author of *Wealth Without Risk* and *Financial Self-Defense*, tells a story about being a guest on "The Today Show" on NBC. On the program, Givens stated his opposition to whole life insurance. A producer from the show later told him that four insurance executives called NBC after the program and threatened to cancel their advertising if Givens appeared again to talk about life insurance. Givens, who's had his own legal and image problems, claims that the executives didn't dispute his position but merely threatened to cancel their advertising.

The industry can reverse its image problems as long as it is willing to take the high road and not distort information. Insurance industry representatives should assist the media in preparing fair and accurate stories instead of being inaccessible.

The industry should be proactive in educating the public about insurance. Wherever possible, the industry should participate in educational programs that explain how insurance works and what its benefits are.

Q-TIPS

Every insurance company should have a policy in place for dealing with the media. One person or department should be responsible for addressing media questions. The media shouldn't be ignored or treated in an adversarial fashion.

CHAPTER 3

Perceptions of Quality and Customer Service

It makes for great television. The investigative journalist is reporting from a tropical resort where insurance agents are mixing business with pleasure. The agents attending the conference are being rewarded by the insurance company for meeting or exceeding sales objectives. The reporter suggests that these goals are achieved by using unethical sales practices.

On another evening, the teaser for the upcoming news report shows an insurance agent driving away from the camera in a Mercedes. There is also footage of the elderly couple he allegedly bilked out of thousands of dollars.

Though the investigative reporter for this television station has a reputation for slash-and-burn reporting tactics, this report is on the money. The agent being investigated has cheated hundreds of consumers.

The agent is guilty of *churning*, a practice frowned upon by insurance regulators and ethical insurers. The agent comes in and makes misrepresentations for the purpose of causing the consumer to drop one policy to buy another. The agent bad-mouths the old policy or the insurer who sold it. By doing so, the agent is able to sell a new policy with its high first-year commission. The consumer loses out, however, because there was probably nothing wrong with the original policy.

Insurance departments across the country find that the practice is pervasive and fine the company millions of dollars. In news reports like this one, the insurance company blames the unethical practice on a few renegade agents.

A former sales manager for that insurance company admits to wrongdoing in an industry publication. He supplied agents with illegal sales material. The former agent claims that the unethical sales practices were encouraged by a corporate culture which valued sales at any cost and a system with little oversight.

This same company is accused of misleading consumers by selling life insurance as a retirement account. The life insurance policies were misrepresented as being superior to an IRA or Keogh plan.

Caught with its hand in the cookie jar, the insurance company finally begins to take action to resolve the problem. It improves the oversight of replacement sales.

The insurance company also implements a quality assurance program to help educate consumers about whether replacing their policy is a good idea. Consumers receive a brochure with questions to ask if they suspect churning. In addition, agents are required to complete a checklist when a client uses money from one policy to buy another.

It's hard to maintain an image of quality after betraying customers. Winning back their trust is very difficult.

Some of the most notorious agents in the scandal were highly respected by the company. They consistently won awards for their salesmanship. Their employer profited from these underhanded sales techniques.

The company had ways to uncover agents who were churning. It was either very lax or deliberately overlooked improper sales practices.

Compliance, ethics, customer service, and quality are tied closely to one another. The new Principles and Code from the American Council of Life Insurance offer concrete advice on improving these areas. The same processes used to improve quality can be used to ensure compliance with the applicable insurance regulations in the state. An insurer's commitment to compliance and ethical market conduct is ongoing, just like the commitment to quality improvement must be.

A large component of customer service is feeling as if you're being dealt with honestly and ethically. When you go to the supermarket and prices are wrong at the register, you begin to suspect that the store is being dishonest. The other possibility is that the supermarket is inept, which won't make you a happy camper either.

Insurance companies sell their products by identifying their customers' needs and satisfying them. If you believe the psychologist Abraham Maslow, we're all driven by physiological and safety needs, as well as love, esteem, and self-actualization. We want a sense of security, and we don't get it from companies that betray us.

Company advertisements appeal to our need for security. If you believe them, you'll get a piece of the rock or will be sheltered by their umbrella. You'll be in good hands, safe and secure.

We all need that feeling of security. Once an insurance company or agent has let you down, it's doubtful you'll feel secure again.

When it comes to selling insurance, quality and customer service are sometimes afterthoughts. Selling the product seems to be the only concern.

Making a sale is occasionally more important than preserving family harmony. After being accused of churning life insurance policies, Metropolitan Life Insurance Company set up centers where customers could file complaints. At one center, a father complained that his son, a MetLife agent, misrepresented the policy.

At a different company, a woman accuses her son-in-law of pressuring her to buy a long-term care policy she didn't want. She didn't know how to say "no" without making him angry.

A large insurance company uses a force of part-time agents who sell life insurance to friends and relatives. The people buying policies are often recruited to become agents, so they can sell to their friends and relatives.

When people are taught how to sell, there's usually a lesson on prospecting. Potential salespeople are told to list all of their friends, relatives, and acquaintances who may be in the market for the product.

While we're not condemning salesmanship, we are concerned about quality service. Agents are in the trenches when it comes to service. They need to be available to provide service and to answer questions. The insurance business shouldn't just be a side business to pick up extra money.

Part-time agents may not provide the service that customers deserve. Sometimes their primary occupation is their number one priority. It's not enough that they have an answering machine to take calls from customers which they may or may not get to after a day at their regular job.

Agents need to be professionals with insurance educational credentials from the Chartered Property Casualty Underwriters and the Chartered Life Underwriters. Obviously, they should be licensed by the state and in compliance with any regulations requiring continuing education.

Whether they're part-time or full-time, every agent should be committed to providing service on a full-time basis. Agents should be rewarded for providing great service, not just for having a great sales record.

Q-TIPS

Every agent and insurance company must work to change the industry's negative image. Unethical sales practices have chipped away at customers' trust in their agents and the industry.

A company's focus should always be on service and satisfying the needs of its customers. By taking care of customers, you'll ultimately take care of the business of selling.

Chapter 4

Image and Marketing

Certainly, the insurance industry's image can't be attributed solely to poor quality, bad customer service, and negative publicity. Even if the industry solved all of those problems, many people would still view insurance as a necessary evil.

Few people get pleasure from buying insurance. They're willing to shell out $20,000 for a car because it gives them pleasure. Buying auto insurance, however, brings no joy to their life.

Many people procrastinate when it comes to purchasing insurance. It is human nature for them to take care of their immediate needs first. Long-term needs like economic security seem to get pushed aside, while people worry about things that must be paid for today.

Talking about insurance makes some people uncomfortable. It means discussing unpleasant topics. Many people hate talking about death and the other perils that life has in store for them.

Generally, there is a mistrust of institutions, even more so when it comes to the insurance industry. It has been embroiled in controversies that have added to the negative perceptions which many people have. When one insurer is accused of wrongdoing, other insurance companies get painted with the same brush.

Differentiation is the process by which one company sets itself apart from another. There are 1,800 life/health companies in the marketplace today. It is tough for consumers to tell them apart.

Differentiation is an important part of an insurer's marketing plan. By bolstering the image of its agency force, a company might set itself apart from the pack. The company touts its agents and their credentials in its marketing campaign. The company will place ads that show the latest additions to its "Million Dollar Round Table."

Companies market by showing differentiation in service. In their ads, they brag that their claims service is quicker and fairer than the other guy's.

Some companies market on the basis of price. If it's cheaper, consumers will come. Some are resistant to change, however. They will maintain the same insurer, even as that company's prices soar above the competition's.

Satisfaction of customers' needs is one of the best marketing tools. Needs selling increases the level of customer satisfaction. The satisfied customer will buy more in the future and will refer prospects to the salesperson.

The American Council of Life Insurance (ACLI) has implemented voluntary standards of market conduct. At the heart of these standards are the Principles of Ethical Market Conduct. A life insurance company subscribing to these principles must abide by them in the sale of individually sold life and annuity products.

Principle 1 obligates the insurer to commit to:

> *Conduct business according to high standards of honesty and fairness and to render that service to its customers which, in the same circumstances, it would apply to or demand for itself.*

Basically, the principle is akin to the golden rule: Do unto others as you would have others do unto you. In a nutshell, the principle simply requires the insurer to be honest, to be fair, and to provide excellent customer service.

While the principle itself is simple, compliance is more involved than just raising your right hand and taking an oath. There are assessment criteria to determine if the company has complied with this ethical principle, as well as with the others. Both the self-assessment and the independent assessment analyze objective criteria to determine if the company is truly in compliance with the Principles and Code.

The insurer must have policies and procedures that are reasonably designed to determine customers' insurable needs or financial objectives in the marketing and sale of its individually sold life and annuity products.

As part of the assessment process, the assessor will determine if there's documentation that identifies how policies and procedures are communicated to distributors, as well as to employees involved in the marketing and sales process.

The insurer must encourage its distributors to use fact-finding tools for determining customers' insurable needs or financial objectives in the marketing and sale of individually sold life and annuity products. Some of these tools include questionnaires, financial plans, customer profiles, and capital needs or financial needs analysis. During the assessment process, the assessor determines if the insurer did, in fact, encourage the use of fact-finding tools by distributors. He or she will look at the resources provided to distributors, as well as formal or informal training in the use of these tools.

The ACLI standards should also improve the quality of insurers' marketing materials. **Principle 4** requires the following:

> *To provide advertising and sales materials that are clear as to purpose and honest and fair as to content.*

The purpose of Principle 4 is to ensure that marketing materials aren't deceptive or misleading. The goal of Principle 4 is that no one receiving these materials should be confused as to their purpose.

Materials presented as part of the sales process must be comprehensible in light of the complexity of the product being sold. To people outside the industry, almost all insurance products are complex. Sales materials should clear up the confusion, not add to it.

Advertising and sales material must be in compliance with the applicable laws and regulations of the state. Illustrations must contain the appropriate disclosure of amounts that are not guaranteed and those that are.

Needs selling, which is encouraged by the new standards from the ACLI, can improve customer satisfaction enormously. The time-honored maxim is that insurance is sold, not bought. During the needs sale, the salesperson helps uncover the client's needs and attempts to match products to meet those needs.

The Principles and Code are major steps toward improving quality, as well as customer service, in the sale of individually sold life and annuity products. Applying the logic of these standards to the sale of other insurance products will help carriers make great strides in improving customer service, quality, ethics, and compliance.

Q-TIPS

Although people will rarely take pleasure from buying insurance, customers should perceive that they are being treated with fairness and honesty. Insurance shouldn't be marketed by confusing the purchaser or misstating the benefits of the product. Good service is much more than a marketing ploy.

CHAPTER 5

Replacement of One Policy with Another

Insurance is a tough sell, and some agents cross over the line to make the sale. They resort to unethical marketing practices to sell policies. A few ignore customers' needs and worry only about selling the product.

Ethical considerations are fine in theory, but agents have bills to pay and quotas to make. When agents need money or must reach sales objectives, the temptation is to move money around.

The customer asks, "What's wrong with the policy I have now?" The answer is probably "nothing."

So why is the salesperson trying to replace it? One possibility is that it's in the agent's interest to replace a policy. On a whole life policy, an agent may collect up to 100 percent of the first year's premium as a commission.

Replacement of one policy with another can be made to sound like a good idea, but it usually isn't. It usually doesn't pay to turn in a whole life policy. The commission has already been paid once. In addition, with whole life insurance, the premium is based upon the insured's age at the time of purchase, so the existing policy may be less expensive.

Frequently, the salesperson will denigrate the policy the customer already has. Ironically, he or she may be speaking negatively about a policy bought from that same individual years earlier.

Because of the high commission and up-front costs, the cash value of a whole life policy doesn't grow much in the first few years. Although the salesperson may provide illustrations that show enormous growth potential, the customer will almost always be worse off for at least a few years.

The agent pushing a replacement policy may try to show how customers get additional coverage for free. Although it may appear as if customers are getting more insurance for little or no money, their existing policy's cash value will be depleted. Furthermore, customers may not need additional life insurance.

In other instances, the agent may convince customers to borrow against or divert dividends from their current policy, so they can buy a new one at what looks to be a reasonable premium. In this fashion, the agent can generate another commission, even though it can undermine the original policy.

Churning is one name for this practice, and it's usually illegal. Most states have a law which dictates that no agent may make a misrepresentation or incomplete comparison of policies which induces a policyholder to buy a policy that insures against similar risks. Replacement is always good for the agent, but rarely benefits the consumer.

Some agents aren't pushing customers to replace one whole life policy with another. Instead, they'll push a term life insurance policy, which has no cash value.

Although many consumer experts recommend term insurance, it isn't necessarily the best choice for someone who's already bought whole life insurance. Term life insurance premiums can increase significantly as you grow older. Whether the agent is selling whole life or term insurance, he or she will gain financially by replacing the current policy.

As mentioned in Chapter 4, the ACLI has implemented voluntary standards of market conduct. A life insurance company subscribing to these principles must abide by them in the sale of individually sold life and annuity products.

Principle 3 presumes that competition is the most effective and efficient means of providing products and services to customers. The obligation of insurers is:

To engage in active and fair competition.

The insurer must develop policies and procedures to ensure that it complies with this principle.

Here's what the insurance company must shoot for, so that it engages in active and fair competition:

- Compliance with applicable state and federal laws fostering fair competition.
- No replacement of life insurance policies and annuity contracts without giving customers the information they need to decide if it's in their best interest.
- A system for monitoring and tracking replacements to see if they comply with company guidelines.
- No bad-mouthing or making misleading statements about competitors by distributors and employees involved in the sales process.

The purpose of this principle is to discourage replacement of one policy with another.

It is clear that companies subscribing to the Principles and Code will no longer tolerate replacement of policies so that the agent can collect a new commission. Insurers will be looking for replacement trends and complaints which show that Principle 3 is being ignored. As part of the assessment process, the assessor will also be seeking documentation that there is full compliance with this principle.

Inappropriate replacement of a policy is a negative marketing practice that insurers should discourage. The company must have guidelines to indicate situations when replacement may not be in the best interest of the customer.

Another negative practice is competitor "bashing." It's not in the spirit of fair play to sell your product by bashing someone else's.

The sale of life insurance should be a positive act, not trashing of the competition. Salespeople should not get down in the gutter and make negative comments about competitors. Principle 3 forces the salesperson to market products in a positive way.

The insurance company must set up a system to supervise the market activities of distributors and employees involved in the sales process. If these people are not complying with the Principles and Code, the insurer can't just throw up its hands and claim that it was unaware of these violations. The insurance company must have policies and procedures in place to make sure this doesn't happen.

Principle 6 requires the insurer:

> *To maintain a system of supervision and review that is reasonably designed to achieve compliance with these Principles of Ethical Market Conduct.*

The company's compliance requirements must be made known to distributors, as well as to all employees involved in the sales process. Compliance training sessions must be conducted for employees involved in the sales process.

There should be policies and procedures in place for monitoring sales practices of employees and distributors. This may be accomplished by random interviews with customers, follow-up surveys, identifying trends, or some other means.

The Principles and Code give the insurer flexibility in setting up a system of supervision and review. To determine whether a distributor is abiding by the standards of ethical market conduct, an insurer might:

- Analyze the persistency of business sold by the distributor.
- Track customer complaints involving a distributor.
- Monitor the level of replacement activity by a distributor.
- Implement measures to detect churning by a distributor.

Nevertheless, an insurer is permitted to develop other methods for meeting the requirements of Principle 6.

Management support is indispensable in establishing an effective market conduct compliance system. Compliance with the Principles and Code should be integrated into the performance evaluation system.

By complying with the Principles and Code, as well as with the applicable insurance regulations, a company will make major strides in improving quality and customer service. Although the Principles and Code only apply to the sale of individual life and annuity products, many of the policies and procedures can be used to monitor the sales practices in other lines of insurance.

There must be oversight of distributors and employees involved in the sales process to ensure that customer service is a top priority in the organization. While replacement of policies isn't usually a problem with other lines of insurance, a carrier should be tracking complaints of any kind against a particular distributor.

When it comes to customer service, the sin of omission can't be taken lightly. The insurer must attempt to uncover mistreatment of customers, not just wait for complaints. Insurance companies can't stand idly by while members of their organization treat customers improperly.

Q-TIPS

When it comes to quality, customer service, compliance, and ethics, an insurance company can't claim ignorance that misconduct occurred. There should be a system in place to ensure that ethical standards are met.

The Principles and Code established by the American Council of Life Insurance provide an ethical framework for insurers to follow, as well as a system for determining if companies are abiding by those standards. Although the Principles and Code are applicable to the sale of individual life and annuity products, any carrier can apply some of the standards to its operation.

Chapter 6

Quality and Selling Can Mix

At too many companies, the salesperson is judged on the quantity of sales, not the quality. The salesperson who replaces existing policies is rewarded, not punished. There are ways, however, to bring quality to the sales process.

Some will argue that over the last few decades, the emphasis in insurance sales courses has shifted from product selling to needs selling. Salespeople are taught to focus on the individual's needs, not just to sell the product. The needs selling philosophy puts the customer's needs at the forefront of the sales process.

The sales process is aimed at addressing the customer's individual needs, not everyone in a particular age group. The goal is satisfying those needs at a price the customer can afford.

In an environment of Total Quality, the salesperson will walk away from a sale where the customer has sufficient coverage. Several insurance salespeople informed us that they rarely walked away from a potential sale. They almost always found that the customer's needs were greater than the amount of insurance already in place.

Insurance regulation is one way to bring quality to the sales process. Laws are passed to prevent sales of more than one Medicare supplemental policy to a customer. Other regulations address whether the sale is appropriate for this particular customer. Almost every state has forms that must be signed to protect the consumer and ensure that the sales process is conducted in an ethical manner. There are also free look periods, so consumers can change their minds and return the policy without charge.

The National Association of Insurance Commissioners has developed "Standards for Marketing." These standards, which have been adopted by many states, prohibit unethical sales practices. *Twisting* is one prohibited act and occurs when the agent knowingly makes a misrepresentation or provides an incomplete or fraudulent comparison of policies or insurers for the purpose of inducing a person to buy a different policy.

These standards also prohibit *high-pressure tactics* and *cold lead advertising*. Cold lead advertising is the failure to conspicuously disclose that its purpose is the solicitation of insurance and that contact will be made by an agent or company.

Insurance regulation is not the key to bringing quality to sales. These are minimum standards, not the end-all for quality.

Sadly, the salesperson who violates the law or engages in unethical practices shouldn't even be reading about quality. He or she has violated every tenet already and is a hopeless case.

Bringing quality to sales is more than just needs selling. It's taking the pain out of the sales process.

Many customers view buying a policy, especially life insurance, in the same negative manner as buying a car. Although they feel in need of insurance, they dread the thought of letting the salesperson into their home. Customers feel the salesperson won't leave until they buy a policy.

It's the same anxiety that people feel when buying a car. They'll spend hours in the showroom and leave too drained to shop around at another dealership. But at least with the car shopping experience, most people know what they want or need. Many do their homework and thoroughly research the car they're buying. Unfortunately, with insurance purchases, many intelligent people don't have the slightest understanding of the product they're buying.

Some would argue that this is why the sales process takes so long. The salesperson needs to educate the consumer and conduct a needs analysis. This is all part of bringing quality to the sales process. Nevertheless, it leaves many people longing for the equivalent of a Saturn dealership with one price and a low-pressure sales pitch.

Perhaps the sales process should be geared to the customer. It should result in satisfying the customer's needs without turning the process into an endurance test. Customers shouldn't feel like they're listening to a never-ending pitch for a time share.

For quality to endure in an organization, the salesperson's needs should be secondary. One salesman used to think about what he'd buy with the commission as a motivational tool. A saleswoman bought a gold-tipped fingernail on each occasion that she made a particularly big sale. When a sale was about to slip through her fingers, she would look down at those nails and step up her sales pitch.

Sales incentives can undermine Total Quality. Companies offer trips and bonuses to salespeople who sell certain insurance products. To qualify, a salesperson may use unethical means to sell the particular insurance product.

It can be argued that the commission structure of certain products encourages unethical sales practices. Regulators and industry groups are questioning traditional methods of life insurance compensation.

The front-loading concept is under fire. Many people believe that unethical market conduct stems from high up-front loading. They fear that this form of compensation encourages agents to replace policies and to sell products that aren't needed.

Opponents of high up-front commissions argue that they provide agents with an incentive to oversell products. This form of commission structure encourages agents to sell the product that pays them the most, not the one that the client actually needs.

Many are pushing for level commission structures. Under these reforms, there won't be a huge payoff in the first year for the agent who sells a new policy. The theory is that level commissions will curtail sales abuses such as inappropriate replacement of policies.

Whether it's life insurance or any other type of policy, agents need to realize that service sells. It leads to customer satisfaction and additional referrals. Providing great service to customers results in more sales.

Q-TIPS

Compliance with insurance regulations is a component of customer service, but not the end-all. Even in a perfect world where insurance companies and agents all comply with every insurance regulation, customer service might still be inadequate.

Ethical conduct is much more than just compliance. One can comply with all of the applicable insurance regulations and still be unethical.

CHAPTER 7

Selling and Customer Service

The John Hancock salesman makes another cold call to a potential customer. A man answers the phone, and the salesman makes his pitch to sell long-term-care insurance.

The salesman asks if the person got the literature he sent on long-term-care insurance. The prospect never got the literature because it was never sent.

In any event, the potential customer isn't interested. He politely asks the salesman to delete his name from the agency's call list.

Instead of deleting the prospect from the agency's call list or even claiming that he will, the salesperson lets his frustration get the better of him. "That's your job, not mine," he says and hangs up the phone.

The salesman moves on to another prospect, while the customer seethes. There is a procedure to shield your name from telephone solicitors across the state, which the customer must initiate, but you can also request that the company making the call not contact you again.

Although the salesperson never left his name, the potential customer checks the yellow pages and calls the John Hancock agency near his home. He describes the situation to the receptionist, who immediately connects him to the manager.

The manager apologizes and assures the caller that his name will be deleted from the agency's call list. He also apologizes for the rudeness of his salesperson. There are four salespeople making calls from his office, and he promises to locate the offender.

Most people aren't cut out to be telephone solicitors. In our minds, telephone solicitation is God's way of saying your own job isn't so bad.

It's not a problem unique to telephone solicitation. Many people feel the same way about junk mail from an insurance company.

Unless a customer is very, very lonely, few enjoy getting these sales calls. Whether they're making love or watching television or both, not many like being interrupted so someone can try to sell them insurance. Or perhaps it's a call from a bullpen filled with solicitors trying to convince senior citizens to join a Medicare HMO.

Cold calls. It's the way many insurance products are sold. The salesperson makes call after call to people who may not have expressed one ounce of interest in the product.

Quality from the salesperson's perspective is a far cry from the potential customer's viewpoint. Even when you're selling insurance by phone, quality and customer-service skills don't have to be abandoned. Whereas some salespeople attempt to bully or scare people into buying insurance, others use the art of persuasion and a gift for gab to sell what may be a very useful product.

You don't have to hit people over the head or make misrepresentations to sell insurance. A Medicare HMO advertises for senior citizens to sell its product. The HMO believes that a more mature salesperson will have success selling the product.

Another Medicare HMO holds informational meetings throughout the community. There's a complimentary breakfast for anyone who'll attend, and the meetings are usually well attended. At consumer fairs, you'll find the HMOs handing out apples or some gadget to draw interest. One HMO offers a free exercise class at the shopping mall to find consumers who might be interested in their product.

On a radio station in South Florida, there's an infomercial for an insurance agency that sells long-term-care insurance. The owner of the agency calls upon one of his salesmen to make comments. He points out frequently that his salesman is 70 years old and has firsthand experience with how important long-term-care insurance is.

The needs sale is another way to sell insurance without undermining quality. As we discussed in Chapter 4, the American Council of Life Insurance has endorsed needs selling as part of its new standards that are designed to improve the industry's image, as well as customer service.

But some companies just don't understand the concept of quality, customer service, and good business. The same John Hancock agency calls back the same customer nine weeks later, using the same pitch. The salesperson asks if the prospect got the literature that was never sent.

The agency obviously had no system for weeding out good prospects from bad. Even though the prospect had shown no interest whatsoever in the product, the agency passed along the name to another salesperson. Occurrences like this demonstrate indifference to customer service, as well as total disregard for their salespeople. This was the agency's idea of a hot prospect for its salesperson to call.

Computer-generated sales calls don't waste the salesperson's time but can be equally infuriating to consumers. A person picks up the phone to hear a recording asking if he needs health insurance. He hangs up the phone.

Almost weekly, he gets the same call. On the fourth occasion, he stays on the line and waits for the opportunity to say his piece. After what seems like minutes, the recording gives him the opportunity to have his name deleted from the company's call list. He leaves his name, number, and explicit instructions to remove his name. A week later, the man gets another computer call that is pitching the same product.

Although quality and target marketing can coexist, there must still be absolute respect for the customer. When the customer shows absolutely no interest in the product, the individual's name should be deleted immediately from the call list. The customer's name shouldn't be passed from salesperson to salesperson in the hope that eventually someone will wear down the prospect.

The hard sell has no place in the insurance marketplace. Insurers who place value on those types of sales methods have a long way to go to improve quality.

Q-TIPS

Insurance can be marketed in clever and ethical ways. The old maxim that insurance is sold, not bought, has no place in an environment of Total Quality.

Needs selling can lead to increased customer satisfaction. Matching the right product with the customer's need is a large step toward customer service.

CHAPTER 8

What Complaints to Insurance Regulators Tell Us about Customer Service

The insurance regulators who deal with consumer complaints and questions are kept very busy. During a recent year, the Pennsylvania Department of Insurance received 30,000 written complaints, 3,000 walk-ins, and 4,000 to 5,000 telephone calls.

Although that may seem like a lot of questions and complaints, it's really not, considering how many insurance transactions take place on any given day. One company alone will deal with hundreds, if not thousands, of customers during a typical day. Nevertheless, the volume of complaints, along with the thousands of customers who never bother to register their dissatisfaction, indicate that the industry has room for improvement in the area of customer service.

Customers can sometimes become alienated over seemingly trivial matters. A consumer calls the insurance department in his state and screams at the person who picks up the phone.

"You bureaucrats have to do something about Blue Cross," he yells into the phone.

"We're trying," the insurance department employee assures him in a calm voice. "What seems to be the problem?" he asks, expecting to hear a horror story about a huge hospital bill that hasn't been paid.

"I've been trying to get through to them for hours," he says angrily. "But I can't talk to a human being. I just keep getting a machine that tells me to punch this number or that number," he complains. "It keeps disconnecting me before I can talk to anyone."

"And what specifically is the problem?" the state employee asks.

"I want to know what hours they're open tomorrow."

After ending their conversation, the insurance department employee tries the number in the phone book for Blue Cross. He finds the phone system is every bit as difficult to navigate as the consumer has said. It is only by calling his contact at the company that he is able to find out the hours, so they can be relayed to the consumer.

The scary part of this situation is that the consumer is pretty sharp but can't get the simplest question answered. Worse yet, Blue Cross offers many policies to elderly consumers who are likely to have even more difficulty with the automated phone system.

The regional manager for the state insurance department has his own health insurance problem that he can't seem to resolve. Although he has group health insurance through the state and an individual medical policy, none of his medical bills are being paid. For a year, whenever a bill is submitted, each insurer rejects it and states that the other policy is primary.

The primary carrier assures him that it is flagging the file so that the problem will not persist. And when the problem continues to occur, the carrier promises to put a second flag on the file. Even after that is done, the same situation occurs.

For a year, he tries to resolve the problem as any consumer would. He never mentions that he is an attorney or threatens legal action. In addition, the regional manager never mentions that he is a regulator who has authority to recommend sanctions against the carrier. Instead, he makes dozens of phone calls, as any consumer would, going up and down the chain of command at both companies, each time believing that the problem has been corrected.

Despite assurances on each occasion that the parties have resolved the problem, both continue to reject the bills whenever they receive them. Exasperated, the regional manager finally mentions that he works for the state agency empowered to regulate insurance companies. And you know what? The problem continues to occur.

It's scary, isn't it? Someone with considerable knowledge of insurance isn't able to resolve a problem. Worse yet, what happens to the people who have no clout, no knowledge of insurance, and a worse problem than this individual's?

You shouldn't have to work for an insurance regulatory body to get customer service. It shouldn't take a complaint to the insurance department in your state either. On too many occasions, it does, however.

The primary carrier was never able to resolve the problem for this individual. The best the company could do was have him submit bills in a personal and confidential envelope to a supervisor who personally processed them. If they were transmitted directly by the provider, the bills were automatically sent to the other carrier.

Many insurers view complaints to insurance regulators as inevitable. Too many of them believe these complaints are prompted exclusively by a negative result. A few insurers even believe that complaints are the result of doing their job right.

Although most insurance companies are in full compliance with the applicable laws in their state, a great many complaints can and should be avoided. Companies that aspire to excellence in customer service will designate individuals or units to handle insurance department complaints. Ideally, those employees should be part of the company's compliance department.

The compliance function of an insurer should be looking for patterns of misconduct, just as insurance departments do. They should attempt to analyze what prompted the complaint and to eliminate the root causes. Employees in the compliance area should make certain that everyone is aware of the applicable insurance regulations and is paying close attention to them.

At more than a few insurers, there are employees who aren't aware that state insurance departments exist. They don't realize that a complaint to insurance regulators is serious and might result in sanctions.

Although there can be severe ramifications as a result of complaints to state regulators, every consumer complaint should be treated seriously, and an effort should be made to find the root cause. Every complaint might eventually wind up on the desk of state regulators.

Q-TIPS

The smallest of problems can result in complaints to insurance regulators. Frustrated customers may also turn in desperation to state legislators, consumer reporters, or the media to voice their displeasure with service at a particular insurer.

Companies need a special unit to respond to insurance department complaints. All employees should receive specific instructions on how to deal with complaints from insurance regulators. Ideally, this unit will be part of, or will report to, the compliance function.

Chapter 9

Quality and the Regulators

Insurers must work hard to meet the expectations of policyholders and other customers. The insurance industry must also satisfy the insurance regulators who monitor the conduct of insurers and agents.

Bankers Insurance Group of St. Petersburg, Florida, came up with an unusual way of dealing with a regulator. The *Sun Sentinel* reported on June 19, 1996 that the insurance company hired a private detective to snoop on a top manager at the Florida Insurance Department.

The company admitted to the surveillance but denied putting an illegal wiretap on the regulator's phone. Bankers' general counsel apologized for the action but offered an excuse. He claimed the insurer was subjected to several insurance department audits and hearings that were politically motivated.

Obviously, Bankers' actions were unconscionable and can never be justified. There were legal actions it could take to challenge the conduct of the Florida Insurance Department.

Nevertheless, insurance regulation is not a perfect science. Many insurance commissioners have political agendas. There are insurance companies that receive favorable treatment from regulators, and that's the antithesis of quality for consumers, as well as the insurers who try their best to abide by the regulations that govern them.

In an ideal world, insurance departments and insurance companies would all be striving for Total Quality. Although this book is about Total Quality in the insurance industry, the regulators should put their house in order too. The things insurance regulators do wrong could be the subject of another book.

The Center for Insurance Research in Cambridge, Massachusetts, conducted a phone survey of insurance department hotlines. The consumer advocacy group concluded that two out of three callers were given misleading information and received little help from the departments contacted.

Being regulated by an inept or politically motivated insurance department can make Total Quality even more difficult to achieve. Customers may question the insurer's action based on a misstatement by the insurance department in that particular state.

In its August 1996 issue, *Money* magazine took the position that state insurance regulation is a joke. The publication claimed that insurers have too much influence over state regulators and lawmakers.

No matter what your position on this issue is, it is clear that insurers must fully comply with insurance regulations. While complaints to regulators don't always mean that the insurance company did anything wrong, they do provide one way of measuring customer satisfaction.

In the manufacturing process, if there are more than three or four defects per million, you need to go back to the drawing board. But what are you shooting for when dealing with customer complaints? Some state insurance departments won't disclose the number of complaints against insurers to consumers. The theory is that big companies will receive more complaints than little companies and that you're doing a disservice by providing this data.

Some state insurance departments keep statistics on the number of complaints compared to the number of policies that the company writes in that state. This would seem to be a way to level the playing field.

Complaints to a state insurance department compared to the number of policies written would seem to be an excellent way to measure quality, but there may be problems in doing so. Perhaps it shouldn't be the number of complaints, but only those complaints where the insurer is found to be in violation of state law.

Realistically, however, insurers do violate state law and get away with it because there is insufficient evidence that a violation occurred. Therefore, data such as this might be suspect. Furthermore, there are many occasions where an insurer has violated no law but has put a customer through the wringer.

There are other problems with using insurance department complaints as a measure of customer satisfaction. A large auto insurer offers some of the best prices on insurance in Pennsylvania. Its service is great, and insureds are treated fairly.

The down side is that the company is quick to cancel if a customer has two or more at-fault accidents in 36 months. For the most part, the carrier is in compliance with the applicable laws pertaining to cancellation, but it takes advantage of every legitimate reason to cancel if a driver seems to be a risk the company doesn't want to continue writing.

Should there be a "yin and yang" when it comes to customer service? The company points to its low rates as evidence that its policy regarding cancellations is legitimate. Its representatives argue that their cancellation practices help keep rates so low.

Another large insurance company puts an ad in *USA Today*, saying it can help drivers save 15 percent on auto insurance. The ad also mentions that the carrier values the relationships it builds with policyholders. The ad states that renewal is offered to over 98 percent of them every year.

At first glance, this statistic would appear to be a good measure for auto insurers to use. Keep track of how many policyholders are offered renewal each year, and benchmark those numbers against industry leaders.

If the percentage is low, it could be argued that underwriters aren't being as selective as they should be in writing business. They aren't doing a very good job of predicting who to insure.

Another good measure is how many drivers do renew their policy every year. A declining percentage might say something about service, although price may be driving the switch to another carrier. Colonial Penn advertises that over 90 percent of its auto insurance customers renew with the company each year.

Q-TIPS

Insurers can learn a great deal about customer service by looking at complaints to state insurance departments. These complaints make excellent training tools. The company that is willing to look at these complaints objectively, and that isn't defensive, will find that many of them might have been avoided.

The absence of complaints is not a sign that customer satisfaction is extremely high. Many consumers are not aware that they can file a complaint with insurance regulators or feel it's futile. In addition, as a general rule, commercial policies are less regulated, and there are likely to be fewer complaints from insureds.

Although looking at insurance department complaints is useful, make sure you're aware of the jurisdictional limitations of regulators in that state. Many state insurance departments do not have jurisdiction over complaints against managed-care organizations and are not empowered to resolve issues of liability that arise under an insurance policy.

CHAPTER 10

Customer Service When the Customer's a Crook

Insurance regulators realize that many complaints are without merit. On many occasions, they rule in favor of the insurer and find that no violation of insurance laws took place. After dealing with certain consumers, insurance departments come to believe that their complaints are not justified.

A state insurance department receives dozens of complaints against insurers from one consumer. He sends the complaints on postcards that have already been used and are absolutely unreadable. He's complained about his HMO, his car insurer, and homeowners—insurance carrier. He can't understand why they're canceling his homeowners policy just because the adjuster found pipe bombs under his bed while investigating a claim.

The consumer with pipe bombs has had 31 claims against his auto insurer at last count. When the regional manager for the insurance department informs the consumer that his office won't respond to illegible complaints, the consumer writes to the insurance commissioner, alleging that the regional manager is in collusion with the insurance industry and that the devil has taken over his body. He alleges that the regional manager accused him of starting the Korean War, where he got his head injury.

Another consumer gives every indication of being an arsonist. The man's homeowners policy was in force for two months when his house burns down from a space heater being used in the middle of summer. He complains to insurance regulators that the bad insurance company isn't paying his claim for damages that are inflated, if not downright fraudulent.

Although he has an attorney, the consumer shows up unannounced in the regulator's office, accompanied by his mother and little daughter. "There will be no Christmas for my little daughter," he says. And it's all because the insurance company won't pay his claim.

Most insurance departments can't or won't force an insurer to pay a claim, especially one that seems to be fraudulent. In most states, regulators can only make certain the insurer is abiding by the unfair insurance practices act or whatever law governs these matters. Usually, those laws only require that a decision be made in a reasonable amount of time.

The unfair claims-settlement practices subchapter to the act lays out the ground rules. In many states, you have 30 days to complete your investigation. If you can't complete the investigation in 30 days, you must notify the claimant at 45-day intervals and give a reasonable written explanation for the delay. Companies are usually given a great deal of leeway if arson is suspected.

On a South Florida radio show, the host complains about his neighbor who filed a phony claim during Hurricane Andrew. The claim for roof damage was paid, even though the receipt submitted was from prior to the storm. Giving customers the benefit of the doubt isn't always the quality solution. Some companies pay the suspicious claim, then pass along those costs to customers.

In Florida, insurance fraud costs each family nearly $1,500 a year in higher premiums. Florida investigators estimate that of $300 million paid out on fire claims each year, $30 million is due to arson.

Improving the quality of investigations at insurers won't eliminate arson. Arsonists are tough to prosecute, and there is a risk of a bad faith action by the insured, as well as punitive damages.

Nevertheless, companies must make the effort and go to the mat on cases where they have evidence that fraud is taking place.

At a minimum, quality means establishing Special Investigation Units (SIU) or Claims Security Units to control the cost of fraud. Many insurers have hired former police officers to help investigate questionable claims.

Even at companies with SIU investigators, employees don't always take fraud seriously. Occasionally, someone calls the switchboard of an insurance company to report a neighbor or relative who's committing fraud. At best, the switchboard operator presses for additional information that the caller usually doesn't have. Frustrated, the caller gives up and lets the perpetrator of insurance fraud go on his merry way.

At too many insurance companies, the people who take most of the calls have little training in dealing with these allegations of fraud. They don't know how to elicit further information or get these callers to the right place. Some insurance companies seem to slough off these callers as kooks or as people with an axe to grind. Ignoring these calls isn't a quality response.

Some companies make the assumption that it's up to the state to investigate insurance fraud. They expect the attorney general or the state insurance department to investigate these allegations. In most states, however, the insurance department fights fraud on the part of insurance companies and insurance agents, not individual perpetrators of insurance fraud.

At a minimum, insurers should cooperate fully with prosecutors and prosecute all perpetrators of insurance fraud. Insurance companies should support organizations dedicated to eliminating insurance fraud, such as the National Insurance Crime Bureau and the International Association of Insurance Fraud Investigation Agencies.

As part of a quality initiative, insurers should improve their ability to fight insurance fraud. Policies and procedures should be implemented, so that employees know how to react to tips that fraud is taking place. All employees should be on board to these policies and procedures and not just handle files

Q-TIPS

Insurance regulators will not force insurers to pay a fraudulent claim. Perpetrators of fraud may try to use the state insurance department to gain leverage and push through a fraudulent claim.

Insurers have an obligation to contest fraudulent claims, so these costs aren't passed along to the vast majority of policyholders who are honest. Clear-cut cases of fraud should be contested vigorously, even if they are expensive to defend. The quality insurance company doesn't buy its way out of a fraudulent lawsuit because it's cheaper than defending it.

Every insurer should create a mechanism for fighting fraud. All anonymous tips regarding insurance fraud should be routed to an employee or unit that can analyze the information and determine whether there is any validity to it.

Chapter 11

Less Than Sympathetic Customers

Insurance regulators don't sympathize with every consumer who complains. Frequently, they see immediately why the insurance company or agent can't or won't help the person.

"That damn insurance company cancelled me," the caller says indignantly. "Just because I was convicted of drunk driving." You don't have to be a member of MADD (Mothers Against Drunk Driving) to lose sympathy quickly.

Coincidentally, there's a call later that day to the same person at the state insurance department. The caller's beef is that the insurance company settled the lawsuit against him, even though he wasn't at fault for the accident. Although the caller admits to being cited for driving under the influence, those charges were thrown out after a protracted court battle.

When the insurance company settled the lawsuit against him, it meant that the caller would lose the $500 collision deductible applied against the repairs on his car. The caller is incensed that he will lose his $500. He mentions casually that the driver of the other car and a passenger died in the accident. His concern, however, was the $500 he lost, not the loss of life.

Several months later, the insurance department employee tries to pacify another angry caller who's been the victim of an "injustice." His auto insurance premium went up $100. The insurance company applied a surcharge to his account because of a serious auto accident. Although the carrier paid out several hundred thousand dollars on the insured's behalf, his only concern is that the premium on his policy went up.

Similarly, insurance regulators may find it hard to empathize with the caller who complains that the insurer cancelled her because she was late with her auto insurance premium. She can't understand it because she's been late every month for the past two years, and they forgave her on those occasions.

Every business deals with customers who may not be worth keeping. The passenger in 4-D is a case in point. The passenger's a frequent flyer, but he's also a frequent drinker. He's on his fourth drink, not to mention the two vodka martinis he knocked down at the airline club.

Sadly, he's not a happy drunk. His voice has escalated in volume during the second hour of the three-hour flight. It's at the point where he's bothering passengers sitting nearby.

The flight attendant refuses his request for another drink, which leads to another tirade from the abusive passenger. If customer service is the end-all, and the customer is always right, then why should anyone be cut off at a bar or at 35,000 feet?

Obviously, customer service has limits in any business. Whether it's the drunk on an airplane or the one who gets behind the wheel of a car, the pursuit of quality and customer service must end at a certain point. For us, the line in the sand must be drawn when a customer acts outside the rules of a civilized society.

An insurance company or any business owes no obligation to the customer who breaks the law or acts in a hostile manner. Threats and obscene language should not be tolerated by employees. Whereas employees should be trained to defuse anger, they need not put themselves in harm's way as they attempt to placate a volatile customer.

Almost every claims department employee can tell a war story about being threatened. The insurance company must put the building guards on alert or call the local police. One claims department employee was stalked at home by an angry insured who believed he was not getting a fair shake. Eventually, charges were brought against the insured.

In Florida, an attorney who handles workers' compensation cases is shot and killed by a claimant. The disturbed individual who shot him felt he had been cheated by the workers' compensation system and later turned the gun on himself.

Claims department employees frequently remark that people get angrier over damage to their cars than they do over personal injuries to themselves or loved ones. It is not uncommon for insurance company employees to be verbally abused and threatened with physical harm because the insurer will only pay for generic parts instead of original factory equipment.

A complaint to the state's insurance department isn't always a breakdown in customer service. Nevertheless, it shouldn't take a complaint to the state's insurance department for someone to get customer service. It happens every day.

In most states, insurance regulators are more than willing to sit down with members of the industry and discuss complaints. Insurers should look at the insurance department in their state as a potential resource for learning about how to satisfy customers. The people who handle consumer complaints for the state can provide a wealth of information about those little things insurers do that drive customers crazy. It may be better than a survey.

Q-TIPS

State insurance departments are usually overlooked by companies that want to improve customer service. State regulators are often willing to provide general information on the types of complaints filed by consumers and what those complaints say about customer service.

Chapter 12

It's a Small World After All

There are problem customers to be sure, but the industry has to worry about cleaning its own house first. To insurance regulators, it's a small world when it comes to identifying those people who fail miserably at customer service or worse. These types would never get a job at Disney World, nor should they be working in the insurance industry.

The state insurance department gets a call at lunchtime from an agent who sells long-term-care policies. He's asking the financial ratings of long-term-care insurers, which the department provides as a courtesy to consumers. The one person who's handling the phone during the lunch hour looks up the rating of five companies and gives them to the agent.

The list continues to grow, however. The list of five companies soon becomes ten. The department employee pages through the thick book from A.M. Best, attempting to find the ratings of companies for the agent.

The insurance department employee hears other lines ringing and asks if he can call the agent back as soon as the other calls are answered. The agent gets belligerent and demands that he be given the other ratings immediately. The insurance department employee provides several more, as phone calls get stacked up on hold. He asks if he can call the agent back in ten minutes with the additional ratings, but the caller won't hear of it.

Finally, the insurance regulator decides he's tolerated enough rudeness. He asks the caller's name so he can call the agent back. Although the agent provides his name, he brusquely tells the insurance department employee not to bother and hangs up the phone without a "thank-you."

The insurance regulator marks down the name and puts it in his desk. It doesn't stay there long. He gets a call the next day from an elderly woman who's in tears. She claims that an agent bullied her into buying a long-term-care policy. The woman states he was rude and told her he wasn't leaving until she bought the policy. She swears he pressured her into buying the insurance.

Surprisingly, or perhaps not, the agent she's complaining about is the rude individual who had called for financial ratings. It's not unexpected that he was too pushy in his dealings with the public. Further investigation showed there were other complaints against him.

Insurance departments look for patterns of complaints. On the local level, it doesn't take a large number of complaints for insurance departments to see a pattern. They'll quickly see that a particular agent or company is causing problems. When an agent is verbally abusive with members of the insurance department, it's not a big shock that he acts inappropriately in his dealings with the public.

Insurers know, or should know, who their problem employees and distributors are. They should be keeping track of complaints and looking for patterns. If it's a small world at the state's insurance department, it must be an even smaller world at each insurer.

Q-TIPS

The people who cause customer-service problems aren't that hard to identify. A company may look for aggressiveness when hiring salespeople, but it shouldn't tolerate abusive behavior toward customers.

In addition to tracking complaints, a company should monitor how many customers take advantage of the "free look" period in the contract. The data may be indicative of a salesperson who pressures customers into buying a policy.

CHAPTER 13

Rate Hikes and Customer Satisfaction

Maintaining customer satisfaction while implementing rate hikes is extremely difficult. Some insurers, however, worsen the problem.

The premium for a long-term-care policy is usually based upon your age when you buy the insurance. The younger you are, the cheaper the premium will be. Agents frequently stress that the premium won't go up, even as you grow older.

The quality agents disclose that the premium can go up if everyone in the same classification's rates go up. The rate hike must also be approved by state insurance regulators. Suffice it to say, few agents highlight this fact in their sales pitch.

The day then comes when a rate hike becomes necessary. It doesn't go up 5–10 percent. The premium goes up by 25 percent for many policyholders.

How do you justify this rate increase to the policyholder? Well, some agents blame the insurance regulators. They approved it. Therefore, they should take the blame.

Some companies write letters expressing regret that a rate increase was absolutely necessary and explaining why this rate increase couldn't be avoided. The letter assures customers that they worked diligently to avoid a much larger increase. Each letter is filled with facts and figures.

Many agents take the initiative and call their clients. They explain that the company is taking bold action so that further increases can be avoided down the road. They rationalize that a large increase is better than being nickled and dimed to death.

Which approach do you like best? Shifting the blame to some faceless governmental entity like the state's insurance department seems to be a cowardly approach.

Where is the consumer left in situations like this? Well, with long-term-care insurance, he or she is left in a bad position. Since the premium is based upon age at the time of purchase, consumers can't just switch companies. If they switch insurers, they'll be paying a premium based upon their current age. They might not be healthy enough to buy a different policy.

Whereas certain rate hikes are unavoidable, injecting quality into the rate-setting process should ensure some level of predictability. The best underwriting process produces a premium level that works for many years and doesn't require large rate increases. Otherwise, the consumer will be left to wonder if the premium was purposefully underpriced to attract customers.

Insurance companies will argue that products like long-term-care policies have a long tail, which means that you won't get a firm grip on the number of claims for many years. But customers don't understand long tails. You certainly wouldn't hang your hat on that explanation in a letter to pacify an angry group of customers.

Customers expect some degree of consistency in pricing. A restaurant won't last long if it charges $10 a meal to develop a following and then doubles the price. The goal is exceeding expectations, not undermining the customer's confidence.

Some suspect that a few insurance companies' underwriting practices are more than just lacking in quality. These companies deliberately distort the underwriting process to market products.

An agent for a large insurance company is at a sales meeting to introduce a new abbreviated-payment life insurance policy. According to the sales manager explaining the plan, the policyholder will be able to discontinue payments after seven years. The agent questions his manager because he doesn't see how the product will work.

Seven years later, the agent is besieged by angry clients who purchased the policy but must still keep paying premiums. The company's position is that interest rates aren't as high as anticipated and that's why the product didn't work. The agent still believes, however, that the policy never would have worked, even if interest rates remained high.

A Utah man writes to *Money* magazine to complain about his experience with vanishing premiums. For a few years after buying a policy, he did not have to pay any premiums to keep the coverage in force. Then, however, he received a premium notice for $5,600. According to the man writing the letter, he confronted his agent, who told him to "read the fine print." And upon reading the fine print, the Utah man learned for the first time that there was no guarantee that the premiums would vanish. He claims the agent never told him that important fact.

The issue of "vanishing premiums" has caused considerable controversy. Many consumers bought life insurance policies after being assured that they would not owe payments after a particular point in time. They learned later that they would still be making payments long after the point in time when they were supposed to be off the hook.

Obviously, customer satisfaction depends on exceeding expectations, not falling short and blaming the problem on events outside of the insurance company's control. Many disappointed consumers have argued that the insurer should have known that dividends would not remain at the high level necessary for premiums to vanish.

A company's reaction to the vanishing premium problem says a great deal about quality and customer service. Many insurers simply ignore the problem. They wait until the customer complains before telling the sad tale that the premium didn't vanish as expected.

Other insurers, however, take a proactive approach. When it becomes apparent that the premium won't vanish, they contact the policyholder and suggest options such as reducing the face value of the policy or increasing the premium amount. While customers are still disappointed, the approach is far better than sticking one's head in the sand and hoping the customer forgets that the premium was going to vanish at a certain point.

When customers buy insurance, most of them expect a long-term relationship with the insurer. Buying insurance isn't fun, so they don't want to go through the process again anytime soon.

With certain types of policies, policyholders expect the premium to rise because of inflation or some mistake on their part. If they crash the car, customers know that their premium may escalate. Policyholders don't expect, however, that premiums will be jacked up significantly for no apparent reason, or that the bills will keep coming after the agent has expressly said they will stop.

Rate hikes and false promises about premiums owed can send consumers looking for another insurance company. It's a situation that neither consumers nor insurers want.

Q-TIPS

In a quality sales presentation, there should be a full explanation of how the abbreviated payment plan works. The agent must provide full disclosure that payments may still be required after the so-called vanish point if dividends are lower than those in the illustration. The salesperson should fully explain the assumptions upon which the illustration is based.

The mere use of the term "vanishing premium" may be misleading. The client should be informed that the premiums themselves don't vanish. Instead, they are being paid through the use of dividends. If dividends don't live up to expectations, nothing will vanish except for the customer's faith in the salesperson and the insurance company.

In addition, the salesperson should fully disclose that the premium for a policy may be increased at some point, if that is a possibility. And if a rate increase becomes necessary, the agent should call customers personally, instead of waiting until they get the bad news from the company. Customer satisfaction might still be saved if the agent is empathetic and can explain why the rate hike was necessary.

If agents go into hiding when there are rate hikes or other problems, their customers should tell them to take a hike.

Chapter 14

Difficult Customers and Customer Service

It's the Pareto Principle at work. You spend 80 percent of your time with a few difficult customers. And Mrs. B is a good example.

The agent and the company he represents have probably lost money on her over the years. The 75-year-old woman has filed numerous auto and homeowners claims. Each one became a major hassle for the company and the agent because Mrs. B squeezed each claim for every penny. She would always attribute preexisting damage to the claims. Because of her persistence and her age, the company would wind up paying her more than she deserved.

On top of that, there were the minor irritations. Mrs. B wanted to pay her auto insurance bill in installments but resented that she was required to pay a small service charge. Mrs. B also wanted the premium break of a higher deductible but screamed bloody murder when she sustained a loss and was charged a higher deductible. All of these considerations were explained thoroughly to her, but the agent is always on the receiving end of a tirade when things go wrong.

Mrs. B always stops by the agency when she needs to pay a bill, since this saves a stamp. Her visits always last an hour as she rants and raves about every conceivable matter to anyone who will listen.

During those visits, Mrs. B asks all sorts of questions. Very few of them deal with insurance. When Mrs. B was having problems with a home remodeler, she came to the agency seeking advice. When an 18-year-old refrigerator broke down, she asked to whom she should write to complain about the "defective" appliance.

Frequently, Mrs. B brings documents that need to be photocopied to the agency. No, it's never an insurance document, just a letter to the light company or some corporation with whom she is feuding. She's always writing nasty letters to someone who has done her wrong.

Mrs. B's a customer, but how far does customer service go? Granted, Mrs. B isn't all there and is a poor soul with no one to talk to except the folks at the agency. Maybe part of customer service is learning how to politely, but firmly, get rid of someone who's in the way.

Customer service is going the extra mile for a customer. The teller at the bank should make a copy of a bill at the customer's request, but the line must be drawn when a stack of documents is brought in for photocopying. It's a bank, not Kinko's.

The lines of customer service have to be drawn sometimes at the point where your business ends. You're in the insurance business. Whereas you have an obligation as a human being to help others in distress, solving all of Mrs. B's perceived problems isn't necessarily part of customer service.

The problem with the Mrs. B's of the world is that they cut deeply into the time you'd spend taking care of other customers. You feel you've provided quality service, but they're back again with a complaint. Worse yet, customers define quality, not you. Customers must perceive that they're getting quality and good service.

It's not going to happen with customers like Mrs. B. They'll never be happy, especially when you cut back on the time you're willing to spend with them.

On any given day, you're a consumer of all types of goods and services. You don't think much about it until something goes wrong. The automatic teller machine doesn't work. An appliance in your home breaks down. Your car won't be ready when promised at the garage.

At your own organization, every employee is in charge of customer service. Employees should be thinking about service all the time, before and after something goes wrong. Unfortunately, no one goes into high gear until after there's a problem, if then.

Sadly, you can shoot for high customer satisfaction in the insurance industry, but you won't always achieve it. Just like Mrs. B, customers have different expectations, and you must meet or exceed them, no matter what they are.

At the barber shop or hair salon, some people love to chat with the person cutting their hair. On the other hand, some of us want to get our haircut quickly and avoid the chitchat. Good customer service is about knowing who wants what.

But just like a bad haircut, results have a lot to do with customer satisfaction. The annuity customer is ticked off because he can't get at the principal without having to pay surrender charges. He doesn't like the choice he made three years ago, or circumstances have changed, and he now needs the principal.

To that customer, it really doesn't matter that the salesperson explained the surrender charges thoroughly when the product was sold. He needs the full amount now. Even when the salesperson points out that 10 percent of the principal is available each year without penalty, the customer still isn't satisfied. He doesn't like the result, so he's dissatisfied with the agent and the company.

Whereas the end result is a factor in customer satisfaction, service can still exceed expectations. Although the agent can't undo the annuity transaction, he can assist the customer in finding other options. The agent can sit down with the individual and discuss alternatives that might meet the customer's needs. Perhaps the customer can borrow against another life insurance policy or find alternative sources of funds. Customers feel the agent is their partner, not just someone who is trying to make a commission.

Even though the result may not be precisely what the customer wants, the agent's actions may result in customer satisfaction.

Q-TIPS

You can't rewrite the policy to satisfy a customer, but you can provide a thorough explanation of its limitations and what alternatives are available, if any. It also helps to communicate why the policy works in that fashion.

Although some customers are more needy than others, you can't let them become overly dependent and consume too much of your organization's resources. Being courteous, but firm, with difficult customers sometimes does the trick.

Chapter 15

World-Class Quality

Despite all of their efforts, Mrs. B will never be thoroughly satisfied with her insurance company or her agent. The problem, however, is that even less demanding customers don't view the insurance industry in a positive manner.

Not too many people run around saying that their insurance company is the best. In advertisements, insurance companies puff out their chests and declare themselves to be the best in the business, but customers aren't praising them to anyone who will listen.

Tooting one's own horn isn't the key to quality. Unfortunately, saying you're the best or world class doesn't necessarily make it so.

In one comedy monologue, comedian Jerry Seinfeld talks about doctors. To hear people talk, everyone's doctor is "the best."

Seinfeld laments that just once he wants to hear someone say that their doctor is the worst, that he graduated at the bottom of his class, or that the man's a butcher.

In the monologue, Seinfeld also observes that patients are always recommending their doctor to someone and to mention their name to the physician. It's almost as if you won't get the right medical advice unless you mention their name. The doctor will give you Tic-Tacs instead of pills unless you mention who referred you.

There's a classic David Letterman bit that indirectly tells us something about quality. The camera follows Letterman around the city, stopping at every greasy spoon restaurant that purports to have the best coffee in town. With a deadpan expression, Letterman asks the proprietor which rating service evaluated his coffee and named it the best in town.

Obviously, Letterman knew that no one from the *Mobil Travel Guide* or the *Zagat Survey* of restaurants sampled the establishment's coffee and certified it to be the best in town. He was simply taking a humorous look at how any eatery can put a sign in the window claiming to have the best coffee.

Insurance companies want their customers to believe they offer the best service in town. They use these claims in the hope of separating their product from the others. To most consumers, all insurance products look alike.

According to Dr. Donald C. Fisher in his book *Measuring Up to the Baldrige*, the world-class organization is one "that produces excellent results in major areas with a sound quality management approach. This organization is totally integrated with a systematic, prevention-based system that is continuously refined through evaluations and improvement cycles."

The American Society for Quality Control offers seven lessons for achieving world-class quality. They are:

1. Know the customer.
2. Get everyone involved.
3. Constantly reevaluate to keep on track.
4. Master the art of relationships.
5. Concentrate on the things you do best.
6. Demonstrate leadership by doing something extra.
7. Learn from the best.

The founder of McDonald's, Ray Kroc, gave even simpler advice. He stressed "quality, service, cleanliness, and value" to his employees.

Many insurance companies could improve just by following these simple lessons. The cleanliness advice doesn't quite fit, but who's going to argue with the success of McDonald's? At another restaurant that is far less successful than McDonald's, the radio advertisement boasts that it offers world-class food served in a world-class atmosphere without paying out-of-this-world prices. Another restaurant promises world-class pizza.

When it comes to insurance, many companies claim to be tops in their field. A self-insurance fund advertises that it offers "World-Class Service."

Many insurance companies seem to think they offer world-class service but don't even come close. They can't seem to deliver on promises of quality and customer service. These carriers disappoint customers at every turn. The only relationship they've mastered is with their advertising agency.

When business turns bad in a particular state, they're quick to stop writing policies there. While there may be good reason to stop writing a particular line of insurance in a state, it's not the way to master the art of relationships. You've walked out on the customer once. The only thing you've demonstrated is that you may walk away from the relationship again if economic circumstances warrant it.

Some relationships may not be worth preserving, as we saw in Chapter 14 with Mrs. B. The agency knows the customer all too well and doesn't like what it sees. On the other hand, customers frequently look at their insurance company and don't like what they see.

Q-TIPS

The term "world class" is more than just an advertising slogan. No organization should cheapen the term by referring to itself as such without going through the benchmarking process. References to quality should come only after a company has thoroughly compared itself to those organizations with the best products, services, and practices. At a world-class organization, the improvement cycles are ongoing.

Chapter 16

Mission Accomplished?

Cartoonist Scott Adams is the author of the best-selling book *The Dilbert Principle.* Adams lampoons corporate life and is no fan of quality initiatives.

Adams offers this tongue-in-cheek advice. He states that if your employees are producing low-quality products that no sane person would buy, the problem can be fixed by holding a meeting to discuss a mission statement. A mission statement is "a long awkward sentence that demonstrates management's inability to think clearly." According to Adams, all good companies have one.

Adams is correct that there's more to improving quality than just formulating a mission statement. Obviously, it's not the cure for a company with pervasive quality problems. Nevertheless, despite Adams' opinion, a mission statement does have value.

Every journey toward quality begins with a good mission statement. The company and every individual in it should know what it is they're striving for. The mission statement makes it clear that all of them come to work each day for a reason other than a paycheck. The mission statement is the "raison d'être." It's the reason for the company existing.

At Atlantic Mutual, here's how the company defined its mission:

- *To provide policyholders with peace of mind against loss from insured events through customized insurance products and risk-management services.*
- *To handle policyholder claims in a prompt, fair, and forthright manner and to resolve just claims against our policyholders on an equitable basis.*
- *To comply with insurance regulations and conduct all our affairs in an ethical manner.*

To meet our responsibilities, we shall:

MAINTAIN financial strength through profitable underwriting and resource allocation and conservative asset management.

IMPLEMENT marketing strategies that differentiate Atlantic Mutual to target customers and SUSTAIN a reputation as a quality specialized property and casualty insurer.

DELIVER products and cost-effective services with our business partners, professional independent agents, and brokers, in select domestic territories and internationally, to meet policyholder needs, all supported by advanced technology.

FOSTER a climate where a diverse employee population develops professionally and personally to better serve our policyholders and enhance performance with our business partners.

At another insurer, the property-casualty claims department had its own mission statement. It declared:

We intend to be the best claims organization in our business. The best claims organization is one which does the best job of

- *Satisfying customers' needs and*
- *Achieving proper claims results by getting full value for each dollar spent.*

Insurance companies and agents don't operate in a vacuum. In addition to dealing with customers, they must address regulators' concerns on a daily basis. There is at least one organization that works to improve those relationships. The Insurance Consumer Affairs Exchange's mission statement reads as follows:

To promote professionalism and shape the standards of behavior in relationships between insurance organizations, regulators, and customers through proactive dialogue, research, communication, and education.

And what is the mission of those regulators? The Pennsylvania Insurance Department's mission statement is a simple one:

To serve Pennsylvania consumers through the fair and efficient regulation of the insurance industry.

The suppliers of services to the insurance industry focus on different issues. The National Council on Compensation Insurance has this objective:

To be the national workers'-compensation information source with a focus on realizing a workers'-compensation system that is equitable to all participants.

And in every organization, mission statements collide with reality. The folks on "Seinfeld" talk about worlds colliding, and they do in the insurance field. References to customer satisfaction in a mission statement conflict with the company's desire to keep claims costs low.

In an ideal world, treating customers fairly will ensure satisfaction, but it doesn't work that way in real life. The insurance company's idea of fair and the customer's don't always coincide. They're not always on the same page when it comes to fairness. Fairness is relative.

The mission statement shouldn't be vague. Too many are. And insurance is vague enough to most people without adding to the problem. It should be simple. Not much about insurance is simple, so at least the mission statement should be.

The mission statement should appeal to stakeholders. Stakeholders are customers, suppliers, and the community. The mission statement should provide direction so that stakeholders aren't charging blindly into the future.

The mission statement is very important for another reason. At every insurance company, too many employees know only *their* jobs. They really can't see the big picture. The mission statement shows them how their duties further the organization's purpose.

Perhaps the former head of USAA said it best. Robert F. McDermott declared that the mission and corporate culture of USAA are, in one word, "service."

THE VISION STATEMENT

The vision statement is a vision at some insurance companies. It's printed on glossy paper and still ends up in the wastebasket.

QualTeam defines the vision as the "picture of the future that defines what we seek to become." The mission is the "specific accomplishments that must be achieved if the vision is to be realized."

According to QualTeam, a vision statement should be brief and memorable, inspiring and challenging, descriptive of an ideal state, and appealing to all stakeholders.

Pacificare of Florida states that certain values keep it on track in achieving its vision. Those include accountability, continuous improvement, customer-driven empowerment, integrity, people, teamwork, and quality. Throughout this book, we will keep reminding readers how important those values are.

Q-TIPS

The mission statement serves as a continuous reminder to all members of the organization. It doesn't just outline what you want to achieve. It also states how you intend to achieve it, and how you'll know if the mission has been accomplished. It is up to top management to formulate the mission statement.

When undertaking a quality initiative, avoid expenditures that employees are likely to view as frivolous and wasteful. Print the vision statement on an item that employees are likely to keep and use, such as a ruler or paperweight.

CHAPTER 17

Customer Disservice: A Tongue-in-Cheek Guide to Alienating Customers

At some companies, their mission seems to be customer disservice, not customer service. It's a dirty job, but some companies insist on doing it. They work hard to alienate the few customers they have left. And if you want your organization to follow in their footsteps to the bankruptcy courts, here's some advice.

The best advice is to talk a good game of quality. In advertisements, prattle on about your company's commitment, dedication, and devotion to quality. Even if your quality initiative is in shambles, don't forget to mention your obsession with quality in every sales promotion.

When it comes to customer disservice, the great ones never renege on commitments without a great excuse. "It's being processed," "We're waiting for the doctor's report," "The adjuster's on her way," "It's in the mail," "You're next on the list," "It's on order," they say with absolute sincerity day after day.

And if you're called on it, deny, deny, and deny some more. "I never said that," "It wasn't me you talked to," or "This is the first I heard of it" are always comforting to a customer who's waited weeks for service.

To avoid Total Quality in your operation, staff it with employees who are empowered to do nothing. These employees are pleasant enough but aren't empowered to do a blessed thing. When a customer arrives with a problem, the employee nods his head sympathetically and promises to bring it to the attention of someone who isn't there at that particular moment.

Those companies that excel at customer disservice are good at benchmarking. They follow the lead of every other insurance company and adopt their programs. They've never had an original idea of their own.

The kings and queens of customer disservice love the form letter, no matter how inappropriate it is. When a customer complains, they send a lovely form letter expressing their deep sorrow but fail to answer the question raised in the letter or to rectify the problem.

You can also use an 800 number to achieve customer disservice. When customers call for service or to resolve problems, the line should always be busy unless you opt to keep them on hold for hours. Be sure to play nice music while they're on hold but interrupt frequently to let them know that someone will be with them shortly.

Companies can afford to antagonize customers at every turn. Otherwise, these customers begin to believe that a company depends on them for its very existence. They also feel a sense of loyalty.

It's the same way with internal customers. Once they're treated with dignity and respect, these employees start to think they have a stake in the company and feel a sense of ownership. And you wouldn't want that if you're striving for customer disservice.

In fact, stay away from the quality improvement process altogether, since it leads to both internal and external customer satisfaction. Otherwise, you'll be taking a proactive approach to solving problems before they happen. Instead, react to complaints and put a bandage on the problem.

If your superiors are forcing you to undertake quality initiatives, don't despair. You still can undermine the quality improvement process and render it useless. Wherever possible, put roadblocks in the way of improvements and hold back resources that are necessary to make changes.

Involve your employees on teams that will work on sweeping changes within the organization. After they come up with improvements, never implement them. If you're not keen on vetoing them yourself, blame your bosses.

Whether it's internal or external customer disservice that you aspire to, don't let quality stand in the way of your mission. By ignoring the process or sabotaging it, you can treat customers as you always have. Until you have no customers left at all.

Q-TIPS

No insurance company purposefully goes out of its way to antagonize customers, but they do so inadvertently. While no one would dispute the value of surveys, supervisors and department heads have inexpensive options that produce immediate results. They can walk over to the closed files and pick out five or ten. In a five-minute phone call, managers can learn if their customers ran into any snafus. The results aren't scientific, but management will learn a great deal about how it is to deal with the company.

They can ask about everything from the phone system to the people who provide service. By calling customers at random, you can get information from those people who may not take the time to fill out a survey.

While an informal approach like this is no substitute for having a quality management system in place, you might view it as a parent-teacher conference. Instead of hearing a status report on your child, you're getting a preliminary look at what your customers are thinking...before it's too late to do anything about it.

Listen to what your customers are saying. Tell them what the company's mission is and see if they feel you achieved it. While you're striving to be a world-class organization, customers may perceive they received second-class service from an organization with no class.

CHAPTER 18

What Do Customers Want?

In Chapter 17, we looked at what customers don't want to hear from their insurance company or agent. The question remains, however, as to what they want from their insurer.

Most people know what they want in a hotel. When it comes to lodging, some just want the basics—a clean room, enough towels, and a bar of soap at a fair price.

Others expect more from a hotel, like toiletries in the bathroom, HBO, and a free breakfast buffet. They want a coffeepot in their room or at least available in the lobby, day or night. Some customers look for a hotel that has a concierge, an exercise room, and dry-cleaning pick-up.

In deciding where to stay, customers can determine which hotels offer these extras and which don't. In many instances, they find that it's the more expensive hotel that's only offering core services, while the cheaper one provides enhancements.

To remain competitive, a lodging chain may begin offering more than just core services. Customers will be entitled to travel awards or fresh-baked cookies at check-in.

As time goes by, these extras become core services that the customer can't live without. Cable television becomes the norm rather than the exception.

When buying insurance, you're not going to get a fresh-baked cookie from anyone. Distinguishing one company from the next won't be easy.

To the typical customer, all insurance policies seem to be alike. In fact, many of the policy forms are standard from one company to the next.

A few insurers go beyond the core services that consumers expect from insurance companies. Liberty Mutual recently began advertising 24-hour roadside assistance to its policyholders. USAA offers more assistance to beneficiaries of life insurance policies than just the death benefit it owes.

Usually, until you need insurance, it's a very intangible product. Policies provide peace-of-mind or a feeling of security, but little else. When you're in an auto accident or your house burns down, insurance becomes a great deal more tangible.

Certainly, some insurance products are more tangible, such as annuities or cash-value life insurance. You get statements showing how much the policy is worth.

Insurance companies can and do eliminate services that customers come to expect. A good customer of one insurance company becomes angry because the carrier changes the payment system. He must now pay semi-annually instead of quarterly. At another company, a fee is added for paying the premium in installments.

From afar, all agents seem to offer the same service. It is only when a problem occurs that the good agent seems a lot different from the bad one.

Every business drops the ball now and then. As a result, they must placate irate customers. In industries other than insurance, the solutions to customer-service problems seem to be easier.

A customer complains to a hotel about a filthy room and a rude front-desk clerk. The hotel chain gives her a certificate for two free nights at any of its properties.

The rental car company receives a complaint letter from a customer. It sends a coupon for $20 off the next rental.

The consumer doesn't like the taste of a new dressing offered by a major food manufacturer. She calls an 800 number and receives coupons good for free food products.

The customer comes for his photographs, and they're not ready yet. When they finally are ready, the developing fee is waived to make up for the inconvenience.

The price comes up incorrectly at a grocery store. To make amends, the store gives the mismarked item to the customer at no cost.

Even the cable company has a remedy when the repair person fails to arrive within the two-hour window. The customer gets $20. No wonder Kramer on "Seinfeld" finally made amends with the cable company after a feud of long duration.

The insurance industry, however, has made few attempts at goodwill gestures. The adjuster fails to show up, and there's no payment forthcoming to the insured. After several calls, the customer is lucky to exact a promise that the adjuster will show up the next day.

There are no coupons in the insurance industry good for $20 off your next premium payment. No one feels so badly about bad service that he tells the insured to skip the next payment.

Granted, the insurance industry is not quite like Kraft Foods or some other company competing for shelf space. Insurance doesn't lend itself to discounts or rebates, especially since giving them will violate most state insurance regulations.

Nevertheless, the industry can learn from its counterparts that sell other products. You should be able to call an 800 number to complain about service. Try it at most insurance companies, and you'll be switched around for hours.

Most insurance company employees find it difficult to cut through their employer's bureaucracy, so customers are in a far worse position. In addition, they don't understand the lingo from P&C to L&H. Their plight is as difficult as asking directions in a foreign country without knowing the language.

Every insurance company needs to have a place where customers can turn for help in dealing with a particular employee or situation. Customers need to know they're not out there by themselves when they aren't getting good service.

Q-TIPS

Each carrier should have an ombudsperson who is knowledgeable about all facets of the operation and all lines of insurance. The individual should be empowered to resolve problems and cut red tape. When the ombudsperson forwards a complaint to the appropriate department, the situation should be given immediate attention.

Problems routed through the ombudsperson should be given top priority, since the customer perceives that the company has not provided good service. If the department generating the complaint has made every effort to satisfy the customer without success, it should provide a thorough explanation to the ombudsperson. If warranted, the ombudsperson should act as an intermediary between the customer and the department. He or she should see if there's an acceptable compromise position.

CHAPTER 19

Stupid Service Tricks

There are hundreds of treatises on how to improve quality and customer service. The authors offer complex theories and complicated advice on how to improve quality. We favor authors who offer simple and straightforward guidance.

Diane Richardson, an assistant editor of the *FC&S Bulletins*, makes the point quite clearly in an article in the June 10, 1996 *National Underwriter*. The headline captures the advice in a nutshell: "'It's The Service, Stupid,' Should Be Industry's Theme."

In the article, Richardson wonders what changed in the industry. When she began her career, most of the people working for the company were intelligent, college-educated, and thoroughly grounded in insurance fundamentals. "Policyholders were to be treated as valued customers, not inconveniences with a fraudulent bent."

Richardson describes several examples of bad service with a common theme. All of the adjusters were trainees. These adjusters ignored the company's rating manual, misinterpreted contractual language, displayed a lack of common sense, and were downright rude.

Having just attended the 1996 Claim Conference sponsored by the Property Loss Research Bureau/Liability Insurance Research

Bureau, Richardson speculates that lack of training is one cause of the problem. She wonders why only senior people seem to attend conferences like this.

Richardson asks why the newer employees aren't there. "Were the topics considered too deep for them? If so, when will the topics not be too in-depth? When they become supervisors or corporate officers themselves? And, in the meantime, where does that leave the insureds who are at the mercy of these adjusters?" she asks readers of the article.

Richardson's questions should cause all of us to wonder about the quality of products and service in the industry. Training is certainly one key ingredient, but money constraints at most insurance companies will cut into the training budget.

Hopefully, the senior people who attend these conferences will bring back information and train their subordinates. Unfortunately, these same people may not have done a very good job training them so far. Otherwise, their subordinates would be more skilled in policy interpretation.

As training budgets are slashed and less experienced employees are hired to save money, something has to give, and usually it's quality. Training is a good investment for every insurance company and agency, no matter how tight budgets are.

Some carriers antagonize employees with their policy on training. Enthusiastic employees find each request for training turned down because of the expense, and eventually, they lose that desire to increase their knowledge.

In fact, some companies penalize employees for their lack of training. On their performance appraisal, employees will be told that they need training in a particular area, yet their request to attend a seminar will be turned down because of budget constraints.

Training leads to better customer service, and that will ultimately make money for the company. Competence is an extremely important dimension of quality service. It's great to have motivated employees, but they also need to have the knowledge and skills to perform the service.

Norman P. Blake, the chairman, president, and CEO of USF&G, said it well:

> *It is my personal belief that we have a "social contract" with our employees to provide training and the proper work environment so they can develop their skills and showcase their talents. In return, our employees have an obligation to make the most of their ability and help to make the difference.*

Q-TIPS

Training doesn't have to be expensive. Experienced employees should be responsible for preparing a training module on a topic of interest. In-house training should be conducted by peers or specialists within the organization.

Carriers need to adopt a train-the-trainer philosophy. Those who attend conferences should lead training sessions back home.

Companies need to use teleconferencing to curb travel expenses. They should purchase training videos or buy the videotapes that are sometimes available from conferences.

There may also be audiotapes on insurance subjects. If these tapes aren't played at formal training sessions, they should be made available to employees so they can listen at their leisure.

Insurance companies should make maximum use of free training put on by outside vendors as a marketing tool. For example, many law firms and expert witnesses conduct free seminars for people in the industry in the hope of attracting new clients.

If employees will be working without supervision, more files should be audited, or more experienced personnel should be hired. If you're not going to train new employees well, you'd better be sure they got their training elsewhere.

The training needs of an organization should not be determined in a vacuum. Seek out employee input, and conduct needs assessment on a regular basis.

CHAPTER 20

Burnout

It's not only the trainees who cause service problems. A few of the veterans seem to have forgotten a great deal about customer service.

It doesn't take a quality expert to see that Kramden has a problem. His co-workers cringe when they hear him on the phone with a customer. It's as if he's doing the caller a favor by looking at his file. Kramden is gruff and almost hostile with every customer.

He's cynical of every person who files a claim. In his mind, none are legitimately injured, and Kramden treats them as such. Although only a small percentage of his case load consists of individuals who have questionable injuries, Kramden's attitude is that they're all faking disability.

There's no doubt he's burned out. Kramden also lacks motivation. Over the years, he's been passed over for promotion and been demoted twice. No one is sure if he was passed over for promotion because of his poor performance or if he became a poor performer after not getting a promotion.

He's been around too long, yet no one knows how to get rid of him. Management hopes he'll retire, but Kramden keeps holding out for a retirement package that never seems to come. He says he can't afford to retire, and it's almost as if that's the customer's fault.

By now, Kramden's co-workers are ready to chip in so he can retire. Much of their day is spent doing cleanup duty of problems he's created.

Whether it's fear of an age discrimination suit or just inertia, management does nothing about its customer-service nightmare. His supervisor gives him mediocre performance reviews, but no one wants to thoroughly document his less than satisfactory behavior toward customers. Management hopes he'll go away, but there will always be a new Kramden to take his place.

If you believe that people want to excel, take a close look at Kramden. He might have felt that way once, but there's no sign he wants to do anything other than skate his way into retirement.

Shifting paradigms is impossible for Kramden. Unless someone holds a gun to his head, he'll handle his job the way he's done it for years. He's totally resistant to changes in the way files are to be handled.

According to Sean F. Mooney, CPCU, a senior vice president at the Insurance Information Institute in New York, claims adjusters see the worst side of their customers. Whereas nine out of ten claims go smoothly, the difficult and unreasonable claimant frustrates the adjuster. After a number of years, adjusters burn out and come to view claim settlements as adversarial. If the customer senses this attitude, the relationship can be poisoned.

According to Mooney, agents and customer-service representatives are also affected by what he calls the 90/10 problem. Ninety percent of the problems come from 10 percent of the customers. Eventually, the problems generated by that 10 percent color the individual's view of every customer.

Your quality initiative will be left smoldering in the ashes if you don't smoke out the Kramdens in your organization. Your customers will be burned up after dealing with burnouts like him. Whether he deals with insureds or claimants, your customers deserve better treatment.

Q-TIPS

Lighting a fire under the burnouts in your organization isn't easy, but the first step is documentation of the performance problems. Customer service shouldn't suffer because management has a problem employee whom no one is able to control.

There should be job rotation wherever possible to keep employees from getting stale and cynical. Even with third-party claims, there should be periodic monitoring of phone conversations for quality purposes. The supervisor should be checking for courtesy, protocol, and attitude.

Chapter 21

Swearing Off Customer Service

"Your employee swore at me," the upset caller says in a trembling voice. "What kind of operation are you running there?" she asks.

The manager apologizes profusely and promises to get to the bottom of the situation. He calls the employee in.

"I talked to the lady," he says. "But I never swore at her. I was agitated, but all I said was 'cheese and rice.'"

"So you're swearing to me that you didn't say 'Jesus Christ,'" the manager reiterates in disbelief. "You only said 'cheese and rice.'"

"That's an expression I use. I never take the Lord's name in vain," the employee says emphatically. "You know how religious I am."

Whether he said it or not, this company swore off customer service for the day. Based on the pattern of complaints against the employee, the odds were good he acted improperly on this occasion. Furthermore, the customer only spoke rationally and courteously during subsequent conversations, which made the manager even more skeptical of his employee. Whether it's "cheese and rice" or a harsher variation, it goes without saying that profanity shouldn't be tolerated by any company.

In getting that quality message across, management often is guilty of the same sins. At meetings, too many managers use profanity to show conviction in the statement they're making.

One claims department manager's use of profanity was contagious. Conversations with him would be laced with profanity. Soon, his subordinates were using foul language.

Maybe it's a cultural problem or a problem in the culture of your organization. Inappropriate language shouldn't be tolerated, no matter where you are.

In many environments, employees become desensitized to inappropriate language. They're used to dickering with body shops or being in other situations where more graphic speech flows freely. You may not be able to change society, but inappropriate speech should not occur in the workplace. It sends a damn bad message about quality, if you'll pardon our French.

Q-TIPS

To eliminate foul language, a department can use a technique that many parents use to curb swearing by their children. Some parents have a rule that children lose a portion of their allowance when they swear. With each use of profanity, a portion of the child's allowance goes into a jar and is put toward a family vacation. Similarly, members of a department may agree to contribute a quarter on each occasion that they swear aloud. The money is then used for a department picnic or outing.

Employees follow the lead of their manager. Supervisors and managers should set an example by not using profanity, no matter how angry they are. The use of inappropriate language in the workplace is an insult to both external and internal customers.

CHAPTER 22

Consistency in Claims Handling

The manager of the claims department is in a bad mood. He got a call last night from a man who fell in a grocery store that the carrier insures. Although there was little evidence of negligence on the part of the grocery store, the claims supervisor made what he believed was a fair offer to the injured man.

Nevertheless, the man was not happy with the offer of settlement. After finding out the name of the manager of the claims department, he looked up the manager's name in the phone book and called him at home. The man kept the claims department manager on the phone for almost an hour, ranting and raving about the extent of his injuries and how inadequate the settlement offer was.

The claims department manager goes directly to his supervisor that morning. "I don't want to hear from this guy again," he advises the supervisor. The manager declines the supervisor's offer to show him the file and repeats the order.

The message is clear: Make the claimant happy, and make him go away. It really doesn't matter that the insured isn't negligent or that the offer is already more than fair.

For the quality message to get across to employees and the public, there has to be consistency. Employees shouldn't be told one day that their job is managing indemnity payout through quality claims handling and then advised to overpay claims on the next.

In a culture that encourages quality, employees don't get mixed messages. They feel that management backs them, even when the customer's a pistol and is hard to deal with in a businesslike fashion.

In addition, in too many organizations, there's one kind of feedback—negative. The employee hears about it when there's a negative survey or a complaint, but compliments are infrequent.

Sure, at the time of a performance appraisal, there will be positive feedback, but it's rare to hear anything good otherwise.

Slip-and-fall accidents account for thousands of claims and lawsuits against businesses each year. In the past, insurers seemed all too willing to simply pay and close them, regardless of liability. Obviously, that wasn't a quality solution to the problem.

It is useful to look at slip-and-fall cases and to see how quality management can improve the situation. Geraldine Prusko, a vice president of CIGNA, offers tips for improving claims handling in a May 13, 1996 article in *National Underwriter*. Although the advice deals specifically with slip-and-falls at a movie theater, the advice can be applied to any setting where slip-and-falls are likely to occur.

According to Prusko, the average person takes 18,000 steps a day, so eventually everyone slips. Prusko suggests forming a multidisciplinary team that includes representatives from loss control and claims, risk management professionals, and the broker. The team can review claims experience and make improvements in the way claims are handled.

The process resulted in the development of a theater kit. The kit includes a script for employees to follow. The theater employee uses it to become a more empathetic listener, a first-aid dispenser, and a claims investigator.

The kit contains an incident form and a disposable camera to take photos of the accident scene. For those people who believe that implementing quality initiatives is too expensive, the cost of the kit was eight dollars. The same process might be used at any business where slip-and-fall accidents are likely to occur. A quality team might work on solutions to prevent accidents at a grocery store and to arrive at better ways of dealing with them after they occur. Maybe the team can help improve the response time when an aisle cleanup is required.

At a discount store one Sunday morning, a customer slips on a puddle of water near the cash registers. When the customer limps over to report the accident, the clerk at the customer-service desk writes down his name and sends someone with a mop to the scene of the accident. No apparent effort is made to investigate the accident. Even a cursory investigation would have documented that two youngsters were squirting each other with water pistols only seconds earlier.

There's no way of knowing if that customer submitted a claim to the discount store's insurance carrier. It is clear, however, that the carrier might help the store improve the process for dealing with accidents.

While insureds can take steps to avoid accidents, claims do occur, and it's incumbent upon the carrier to develop standards for handling them. Organizations have implemented BEST practices to standardize the way claims are handled and to improve file quality.

For example, two-day meaningful contact with the claimant might become the standard. Adjusters must speak with the individual in 48 hours or explain why they didn't. Files are routinely audited to make certain the adjusters achieve meaningful contact on a high percentage of their files.

A 48-hour-or-less contact requirement has been shown to reduce claim payout. Prompt contact increases the possibility that the claimant will deal directly with the insurance company instead of running to an attorney. The cost of resolving the claim is likely to be lower without an attorney being involved.

Although this is a litigious society, not every injured party gets a lawyer out of greed. Some retain a lawyer as a result of bad service. Someone from the insurance company doesn't return a call, so the person hires a lawyer who won't return phone calls. Prompt contact makes it less likely that claimants will engage counsel.

Implementing timely contact requirements does not guarantee that insurance company adjusters will abide by the standards. The claim representative may simply make a notation in the file that she called the claimant and that there was no answer. Others may send a postcard asking the claimant to contact them and then put the file at the bottom of the pile.

Ideally, however, quality standards produce quality results. The adjuster will genuinely attempt to contact the individual, even trying from home in the evening or on the weekend.

In addition, there is value in creating standards to use in evaluating and settling cases. For example, the organization may create an evaluation form to place a fair settlement value on claims and lawsuits. If an honest effort is made to use the form, there will be consistency in settling cases throughout the department. One drawback is that an adjuster may fudge the sheet to justify whatever settlement is made.

Changes in claim-handling philosophy can also undermine quality. During one month, management is encouraging its employees to be tough but fair. If it's clearly a case where there's no liability on the insured's part, contest it. A month later, however, the carrier decides its legal expenses are rising, so the employee is pushed to settle cases, even those where liability is remote at best.

Many carriers have reputations for holding fire sales at the end of the year. Plaintiff attorneys know that claims employees will be under pressure to settle cases and will increase their "absolutely final offer" by thousands of dollars.

Defense attorneys for the insurance industry sometimes send the wrong message about quality and customer service. Up until the eve of trial, they discourage settlement, saying the case is a sure winner. Some suspect they hope to bleed the last cent out of the case before recommending settlement at the last minute.

An important part of achieving claims excellence is improving the quality of defense firms that handle cases. Insurers want a quality defense while keeping expenses under control. Quality claim departments will use litigation management techniques, as well as attorney evaluation forms. They will also explore alternative dispute-resolution mechanisms and creative ways of controlling legal expenses, such as a prearranged fee for handling a particular case.

Innovation is the key to achieving excellence, not the same tired old ways of doing business in the claims department. At Reliance National, the professional liability and financial specialty lines have integrated their functions. When the vice president of underwriting for architects' and engineers' professional liability left the company, his job was given to someone in claims.

Traditionally, there has been interaction between the claims and underwriting departments. Reliance National carried it a step further and integrated those functions. It lets the company get a handle on claims trends quickly so the insurer can adjust the way business is written.

USAA has initiated a "Claimant as a Customer" program. USAA's practice is to contact claimants within 24 hours of the accident. Claim representatives are available 24 hours per day, seven days a week. By making early contact with claimants, USAA is able to resolve claims faster. The company is able to establish a rapport with the claimant and minimize the possibility of litigation.

USAA points to a report from the Insurance Research Council, which shows that half of all claimants who retain a lawyer do so within a week of the accident. The report found that claims where an attorney is involved take longer to settle.

Claimants are owed the same level of customer service as insureds. There should not be two different standards of service applied. Nevertheless, many carriers seem less concerned about complaints from third parties than from insureds. They're all customers and deserve the same high level of service.

Q-TIPS

The claim-handling philosophy shouldn't change drastically from month to month or depend on how hostile the claimant is. While there should be flexibility in the claim-handling process and a willingness to compromise, there should also be consistency in the way cases are resolved.

Standardizing the way claims are handled leads to improved quality and customer service. The people who do the work should be instrumental in developing standards.

CHAPTER 23

Workers' Compensation Best Practices

Handling a slip-and-fall claim is much different from handling a workers' compensation claim. One difference is that states usually impose strict time limits on deciding whether a workers' compensation claim is compensable. The time limits for resolving other claims are usually defined by the unfair insurance practices act in that state.

As mentioned previously, state standards on claims handling do not define quality. Compliance with existing statutes and regulations is just the starting point on the journey to quality.

In an ideal world, resolving workers' compensation claims wouldn't be so tough. All of the parties would be united in their desire for a workers' compensation system that is equitable to all participants. Workers would want to return to their jobs as soon as they feel better. Employers would patiently await the injured workers' return.

But real life isn't so easy. Many workers aren't so anxious to return to their jobs. They may be making more money on workers' compensation than they would by working. Employers, meanwhile, are normally chomping at the bit for the employee to return to work. Frequently, there will be no replacement for the injured worker.

Costs can be lowered and claims administration can be improved through the quality improvement process. The process can be used to develop quality review forms.

Although they are subjective to a certain extent, quality review forms can lead to continuous improvement. The forms must be specifically geared to the nature of the work. The theory is that if you dot the "i's" and cross the "t's," you'll achieve a quality result.

Let's take the workers' compensation claims process. The form would look at the quality of the investigation. Was there timely contact with the insured and the claimant? Was the investigation thorough and well documented?

Were subrogation issues addressed? Was the correct decision on compensability made? If a second injury fund exists in the state, did the claim handler consider that issue?

The quality review form should certainly address whether the correct coverage decision was made. Quality really is moot if your company shouldn't have been covering the claim in the first place. At a minimum, you'd also want to double-check whether the rate of compensation is correct.

The form should address claims and litigation management. Did the claim handler put the file in a drawer for months or let the attorney run wild with it? Were legal bills cost controlled or just initialed for payment? Could litigation have been avoided?

The review form would also look at medical management. The claims handler may have forwarded the file to a rehab nurse and forgotten about it. Were the bills paid directly related to the accident and were they paid promptly? Were the bills cost controlled?

Reserves should be looked at too. Were they calculated properly, or were they just a wild guess? If the account carries special instructions, was the insured notified on claims with high reserves? Did the claims handler make a rational attempt to predict the claim's value or just increase the reserve each time that payments were made?

Every state has requirements pertaining to forms that must be filed. The quality review form should address whether all of those requirements were complied with by the file handler.

Obviously, these are just a few of the elements that should be incorporated in preparing a quality review form for workers' compensation. The form would be quite different for another line of insurance.

Quality in the workers' compensation claims area will not result just from the use of a form. Carriers must use quality management to arrive at solutions for their own organization. As a starting point, employees should make use of:

- Benchmarking
- Quality measures
- Cross-functional teams
- Self-directed work teams

They should interview all the parties and look at work flow. In doing so, they may find there are bottlenecks in the system.

Teams may find it worthwhile to:

- Implement BEST practices like two-day meaningful contact, not just a mailed postcard.
- Use automated letters.
- Educate employers about the need to report claims quickly.
- Fax notices of claims to outside adjusters instead of using the mail.
- Develop quality measures such as ratio of medical, indemnity, and expenses to lost time; average claim cost by body part; and percentage of claims in litigation.

One major carrier came up with these specific changes to improve the quality of workers' compensation claims handling:

- A post office box for each department instead of letting correspondence go through the mailroom, which had trouble distinguishing a workers' compensation bill from medical submitted on an auto accident or health insurance claim.

- Paperless files.
- Structured settlements to resolve claims.
- Preferred provider organizations to contain medical costs.

While they aren't foolproof, quality review forms can help improve customer service. Every organization should develop forms that are tailored to its business activities. Managers should be periodically reviewing files as a way to develop a baseline and to measure improvement.

Almost every function should be able to develop objective criteria for evaluating its work product. The people who do the work should develop standards that must be met in performing their duties.

At a minimum, the form should measure how long the response time was in comparison to a standard. In formulating standards, the team should determine how long it should take from the date of contact to the time a decision is made. How long should it take to resolve the situation? Once again, there should be a standard to help measure the timeliness of the claim-handler's actions.

The review form will look, too, at the quality of the decision made on the case. The reviewer must decide if all of the legal issues were addressed and if all factual questions were investigated.

Obviously, the review will still be subjective, to a certain extent. Someone will be judging the quality of the investigation, and that judgment won't necessarily be objective.

In theory, meeting or exceeding the standards set forth on the quality review form will result in a happy customer. Unfortunately, however, doing everything by the book won't always result in customer satisfaction. The employee whose injury isn't compensable is unlikely to rave about the carrier's customer service. The standard, or even a higher standard of performance, won't please everyone.

Although you can't please everyone, especially someone who may be in an adversarial position to your company, you can raise the level of performance in dealing with workers' compensation claims.

Q-TIPS

Because time is of the essence in workers' compensation cases, every effort should be made to reduce delays in the claim-handling process.

Quality review forms and quality measures should be used as educational tools. The results should be discussed with employees in a positive way. The results of reviews should be used as a baseline, and improvement should be measured.

CHAPTER 24

BEST Practices for Agents

BEST practices aren't just for the claims department. The Feitelberg Agency in Falls River, Massachusetts, participated in a pilot program with the Westinghouse Productivity Center. The improvements led to an increase in profits, as well as productivity, which was measured by revenue per employee.

The participants in the project used 12 conditions of excellence, which closely followed the Malcolm Baldrige National Quality Award criteria. By participating in the project, the agency was able to improve billing statements that were confusing and frustrating to customers. The agency also dealt with other problems like unbalanced workloads and the lack of a formal method for acting on customer feedback.

But not every agency cares about making improvements. Six hundred miles away from Massachusetts, an agent seems bent on destroying his agency's productivity. He's managed to alienate two customers over a three-dollar charge.

Only weeks earlier, a client had recommended the agent to his aunt who is in the market for a renters' policy. The agent seems nice enough, and the woman writes a check for the cost of the policy.

The agent tells her that if she writes a check for the full amount, the woman can avoid the three-dollar service charge for paying in installments. The woman writes a check for the full amount, $128, and the agent writes a binder for the coverage.

A few weeks later, she receives a bill for $23 more. The agent priced the coverage incorrectly, which resulted in an increase of $20. And the three-dollar charge came about because the woman didn't pay the bill in full.

In many cases, you can boil customer service down to who caused the problem. If it's the insurance company or the insurance agent, they had better fix it without further inconveniencing the customer.

In this instance, the agent made an innocent mistake. Since this is only a quote, the customer should be responsible for the additional charge. Nevertheless, imposing the three-dollar charge for failing to pay the bill in full is ludicrous. The customer had no way of knowing that she was paying less than the full amount.

This particular customer became incensed that an additional three dollars was charged to her bill. She refused to pay it. In doing so on principle, she risked the company cancelling her. In some states, failing to pay the full bill, despite these circumstances, may result in a cancellation that cannot be overturned by insurance regulators.

Aside from the cancellation issue, the insurance agent managed to sour the relationship in weeks...not because of the innocent mistake but because he refused to waive the three-dollar charge or pay it himself. The agent undermined what might have been a profitable relationship with this customer.

In doing so, the agent also lost the nephew's business. The nephew was embarrassed that someone he recommended to his aunt gave her a hard time. He had several policies with the agency, not just a renters' policy.

The pursuit of excellence at an agency isn't just avoiding the initial mistake that causes hard feelings with a client. It's knowing how to correct unavoidable mistakes, while maintaining customer satisfaction.

In any business, increased profitability comes from improvements in quality and customer service, not from bleeding customers for the last nickel. That point was well illustrated by a recent letter to Click and Clack, the syndicated columnists who write about cars.

The author of the letter went for a Mr. Goodwrench Oil Change Special that was to cost $24. When the dealership called to inform him that his car was ready, the service technician casually mentioned that they had topped off his windshield wiper fluid. The customer was pleased, since he had not asked for this service and thought the dealership had gone the extra mile for its customer.

When he picked up the car, the customer found that the bill contained an additional charge of $6.36 for windshield wiper fluid. Although the customer was charged over six dollars for the service, the actual cost of the fluid added to his car was about 25 cents.

Both columnists were angry at the dealership. Tom Magliozzi, one of the two brothers who write the column, was amazed at how the dealership managed to change a delighted customer "into a guy who writes a nasty letter that the whole country is going to read." Instead of going back to Mr. Goodwrench every three months for an oil change and other service, the dealership lost a customer over $6.36. The customer will also tell his friends about the negative experience.

The lessons from this letter apply to any business. Relationships require give-and-take among the parties, and one side cannot exploit the other. Relationships are long-term propositions. Those who gouge customers for short-term profits won't survive.

In choosing winners of the Baldrige Award, the judges and examiners look for ten core values and concepts. Among others, there must be customer-driven quality and continuous improvement. A company applying for the award must also have a long-range outlook and a long-term commitment to customers.

Whether it's three dollars, $6.36, or a lot more money, shortsighted treatment of customers drives quality out of an organization. That business and its customer will never get together again, much like oil and water.

Q-TIPS

Mastering the art of relationships requires flexibility and a willingness to admit mistakes. An insurer or agency should think long and hard about the implications before drawing a line in the sand with a customer. Even if you can't accept that customers are always right, they are certainly owed the benefit of the doubt.

The cost of acquiring new customers is expensive. Squeezing a few extra dollars out of a customer never makes sense financially and will jeopardize any hope for a long-term relationship.

CHAPTER 25

Timeliness and Customer Service

Most of us have too many things to do and too little time. In view of that fact, the quickest way to lose customers is to be indifferent to the value of their time.

We mentioned earlier that the cable companies have taken a small step toward recognition that customers don't need to waste their precious time waiting for service. Some cable companies pay customers $20 if their service technician doesn't show up at the appropriate time. Although $20 may not be adequate compensation, it is a step in the right direction.

There are many insurance companies that don't recognize the value of their customers' time. The insured is told that the adjuster will drop by tomorrow, but no one can even guarantee whether that will be in the morning or afternoon.

In another era, there usually was someone at home during daylight hours who would be available when the insurance company representative deigned to drop by. Today, however, that would be the exception rather than the rule.

The situation is just as bad with return phone calls from some insurance companies. The customer waits by the phone for the call, afraid to leave the house or even to step outside to get the mail. If the caller actually speaks with a human being and not a machine, it's rare for anyone to commit to a time when the call will be returned.

Disrespect for the customer's time is not limited to the insurance industry. At the Planet Hollywood in New York, we stood in a long line that stretched past the Russian Tea Room. The Planet Hollywood employee who controlled the line assured us that it would take 40 minutes to get a table.

After an hour, the line had hardly moved, so we asked again. Now we were told it would be about a half-hour wait for a table.

As we drew closer to the door after 45 more minutes, we asked a different employee for his estimate of the potential wait. He indicated that we now had a 30-minute wait, plus another hour once we reached the inside of the restaurant.

Perhaps we tourists get what we deserve. We wait in line at Planet Hollywood, Disney World, or wherever, not realizing that whatever awaits us can never make up for the amount of time it took to get there.

Some businesses try to keep customers informed. There will be signs at various points in the line, indicating how much of a wait still remains.

The customer-service line at a mail order house has the right idea but can't pull it off. The automated phone system tells callers how long they'll wait…sort of. They're told that the wait will be more than 90 seconds.

Well, the wait is 90 seconds and then some. After 20 minutes on hold, the potential customer gives up.

Customers wait for service at insurance companies, not really knowing when it will happen. Some companies take positive steps to keep customers informed.

One health insurer has a monitor in its waiting room telling walk-in customers when a representative will be available. And based on experience, the actual wait is never longer than the time promised.

Nevertheless, some insurance companies take the position that their service is worth waiting for and that they'll get to it eventually. These insurers must be prodded into providing service. Their customers waste valuable time waiting for service, as well as the time they spend trying to get the insurance company to provide service.

It's doubtful that the customer waited very long when the policy was sold. On the sales side of the insurance industry, appointments are made, and a salesperson arrives on schedule. Otherwise, it's doubtful that the sale would be made. Apparently, many in the industry believe that you need not be as punctual when there's no financial incentive for being on time.

Indirectly, however, there is an enormous financial incentive to provide timely and effective service. The companies that do so will benefit financially.

Some insurers know that consumers expect bad service from the industry and hope to change that perception. Progressive Insurance promises in an advertisement to change your expectations about auto insurance. In describing its Immediate Response program, Progressive boasts that 70 percent of damaged vehicles are inspected in one day and most collision damage claims are wrapped up in a week.

If that promise is kept, it will alter the expectations of thousands of customers. Those customers expect nothing but aggravation in dealing with their insurance company, and frequently that is in fact the case.

Q-TIPS

Wherever possible, give customers an estimate of the time it will take for service. Estimates like more than 90 seconds are worthless.

Implement a policy of returning phone calls within an hour or less. In addition to voice mail, supply employees with pagers so they can return phone calls immediately.

When an insurance company has little or no regard for the customer's time, it's time for the customer to change companies.

CHAPTER 26

Accessibility and Customer Service

Because people lead such busy lives, they are attracted to businesses that offer service at unusual hours. These individuals have very little time during normal business hours to deal with personal issues.

Every now and then, you'll see it in an advertisement for a plumber or some other business. The ad proudly proclaims that you can call 24 hours a day for service.

Those of us who are naive call those numbers thinking we're going to hear a live voice on the other end of the line. More than likely, however, an answering machine picks up the call, or the answering service promises to leave a message for the person who will return the call in the morning, if you're lucky.

The customer calls the 24-hour banking center. The person who answers the phone is lovely. Although there is an obvious error in an account, she does not have authority to correct it. Even after conferring with her supervisor, the bank representative must hand off the problem to a daytime supervisor who never calls the customer back.

In attempting to find out where his problem stands, the customer calls during daylight banking hours. No one knows which supervisor was assigned to correct the problem. When the customer calls back that night, he finds the person he talked to previously only works three days per week. Her desk is locked, and no one knows whom she talked with about the customer's problem.

The customer is getting shortchanged on service and money. He had increased the dollar value of his accounts to qualify for additional services with the bank. Instead, he's now spent several hours on the phone attempting to correct a $22 error.

Promises of service should not be illusory. An insurance company number that promises 24-hour claims reporting shouldn't just have an answering machine to which the customer can pour out his tale of woe. If promised, it should be staffed by knowledgeable people who can offer advice and put the wheels in motion toward resolving the claim.

GEICO Direct advertises "complete 24-hour service," not just an answering machine. At any hour, you can get a quote or update your policy. The advertisement promises that the company will process your claim, day or night. Certainly, that type of service will be particularly attractive to harried customers who have very little free time.

GEICO's advertisements play heavily on the advantages of dealing by phone with the company at all hours. Customers are able to do business in their bathrobes. Furthermore, there's no need to clean the house because the agent's coming over.

In many infomercials and commercials, you'll hear "Operators are standing by to take your call." Businesses don't want you to waste a second of your precious time if you're ready to buy. After all, you might change your mind about the purchase.

It would be refreshing to see insurance companies standing by at all hours to provide service. Unfortunately, many aren't even standing by during regular business hours to provide service. While the customer waits impatiently on hold, the recording plays over and over, "Your call is very important to us. Please don't hang up. We'll be with you in a moment."

After hearing the hollow promise for the eighteenth time, it dawns on the customer that quality service isn't that important to the company. Service should be easily accessible. It should be a phone call away and not an interminable one.

Q-TIPS

Customers don't expect to get service outside of normal business hours unless the company creates that expectation. They do, however, expect insurers to make an effort to accommodate their schedules.

Insurance companies can extend hours to provide service while increasing employee satisfaction. By providing flexible work hours for employees, the insurer can have employees available at times other than normal business hours.

In their job descriptions, it should be made clear to certain employees that they do have obligations beyond the normal work hours. Claims representatives, for example, should be made aware that there may be occasional evenings or weekends when they'll need to call customers. Similarly, appraisers may be required to make special appointments for customers who aren't available during the normal working day.

An insurer should never promise access to the company at unusual hours unless it can deliver on that pledge. Voice mail or answering machines do not constitute service, unless someone will follow up on the call in a reasonable amount of time.

If there are peak times when phone calls stack up on hold, the insurer or agency should allocate its resources accordingly. Banks, for example, hire tellers who only work at peak times of the day or week. These nontraditional work hours appeal to many potential employees.

CHAPTER 27

Waiting for Service

Instead of seeing businesses standing by at all hours to provide service, customers frequently seem to be standing by at all hours waiting for service.

A visitor to South Florida breaks down in what appears to be a bad area. It's near Miami, where several foreign tourists were murdered in recent months.

He calls the national number for road service from a phone booth where the windows have been broken by vandals. Despite calling four times, he waits four hours for a tow truck to come.

In his home town, the motor club has been very responsive when he's called in an emergency situation. After he returns home, he complains about the four-hour delay in getting service and is told apologetically that his local motor club can't control the quality of garages outside its territory. At the time his membership comes up for renewal, the member joins a different motor club.

Many motorists who are in their cars a great deal feel better if they belong to a motor club. The coverage provides them with peace-of-mind, much like insurance does. Customers believe that no matter where they are, the motor club will help get their car back on the road.

Lack of control is never an excuse for bad service. There must be a mechanism in place to control the quality of service at every level and in every part of the country. Customers shouldn't be left to feel that the operation in certain areas is being run by loose cannons who answer to no one in the organization.

Yet many businesses seem to run that way. Even the supervisors at this particular motor club seem to have no authority to order a tow truck to help the customer. They are only able to apologize profusely and make promises that aren't kept by the people who do the work.

A family ships an automobile to Phoenix from New York. Weeks later, the person who answers the phone at the shipping company has no idea where the truck is in relation to Phoenix. Even the intervention of the Interstate Commerce Commission can't determine where the vehicle is and when it will arrive in Phoenix. The only defense offered by the shipping company is that it has no control over the people who physically transport the car from place to place.

To insurance company customers, the situation may seem the same at times. The customer calls the person who makes the assignment to appraisers, but days go by with no contact. The undrivable car sits in the individual's garage, waiting to be appraised. Each phone call from the customer is answered by the same courteous person who tries every tool at hand to obtain service for the customer, but fails.

Agents find themselves in the same uncomfortable position. They are in the hot seat with customers, trying to explain why the carrier has not fulfilled its obligation under the policy. Even though the agent is applying pressure on the company, service is still not performed in a timely manner. The agent apologizes to the customer, saying that the situation is out of her control.

In some organizations, control is like Big Brother, but in a more positive sense. Phone calls with customers are monitored for quality assurance purposes. But where employees have enormous freedom, quality control may be nonexistent.

Whereas the inside adjuster can be monitored relatively closely, there are many situations where claim handlers are given a great deal of freedom. They work out of their homes and only show up in the office for meetings on a monthly basis. Their new notices are faxed to them or mailed.

Sometimes insurance companies must rely on people they don't even know to provide service. Many carriers must retain an independent adjusting firm when faced with a claim in a particular area. They may never have used the firm before, nor is it likely they'll need the firm again.

Lack of control or unfamiliarity with the firm is no excuse for bad service or poor quality. There must be control over outside vendors like independent adjusters. Carriers must exercise an extreme amount of care in selecting these vendors.

Even after they've been selected, it's still important to monitor the work of vendors to make sure they're meeting your expectations. It is helpful to follow up with customers to be sure they were treated well by the vendor. Ideally, there will eventually be no need to closely monitor the firm.

The president of a large insurance agency tells the story of his favorite independent adjusting firm. Over the years, they've provided excellent service to his customers, and he repays them with his loyalty. When he's looking through the invoices, the president sees a small payment to a different firm. Although the check is under $200, he wants to know why. He calls the employee who assigned the file into his office and demands an explanation. The employee gets the message that this is the vendor of choice and acts accordingly from then on.

The president of that agency is wise to stick with an adjusting firm he trusts. An adjusting firm can do far more damage than just antagonizing a customer.

According to the sketchy facts in a wire service story on October 11, 1996, an adjusting company cost an insurer more than $1 million.

It was reported that the adjusting company delayed sending $10,000 to a man the insured rear-ended. When the man whose car was hit didn't receive payment from the adjusting company within 50 days, he sued the insured and won $600,000. Subsequently, the insured sued his own insurance company and won more than $1 million. The jury decided that the insurance company acted in bad faith by not settling the original case for $10,000 when it had the chance.

Obviously, there's a lot more to this story than what appeared in the six-paragraph news article. Since the accident occurred in 1986 and litigation ensued for years, no one except the lawyers is likely to be clear about what transpired.

One thing is clear when federal agencies choose vendors. There's a lot of paperwork involved. The process, however, doesn't necessarily result in good service.

Flood insurance is a federal program. When a flood occurs, the Federal Emergency Management Agency sends in independent adjusters to settle claims. During one Pennsylvania flood, an adjusting company was sent in from Texas. And in the opinion of many people who filed claims, those outside adjusters didn't provide good service. They set up shop in a hotel room and calls could not get through to them.

As you might expect, in any disaster situation, many people will not get the service they expect during a very trying situation. When the local motor club gets 100,000 calls during a bitter January from members stuck in snow banks or with frozen fuel lines, the service won't be as good as it is on a balmy July evening—unless, of course, you're that stranded motorist in South Florida who got bad service on one of the prettiest nights of the year.

During Hurricane Andrew, insurance companies sent in thousands of adjusters from offices across the country, but they still were shorthanded in coping with the disaster. They were also plagued by a shortage of building supplies and prices that escalated because of the shortfall.

Sometimes scarce resources aren't allocated properly. Adjusters are sent to the stricken area, leaving too few to handle the routine claims. Whereas certain areas are getting good service, the folks back home are suffering.

Allocating resources in emergency situations is an important part of quality. While you'll never be able to perfectly predict where help is needed most, the key is flexibility. How do you react once you realize your resources aren't allocated properly?

The federal government sets up flood relief centers based on political considerations, and sometimes leaves them open even if the need doesn't exist. An insurance company doesn't have to run that way.

Some organizations seem to be impotent when it comes to providing service. They'd love to help but can't seem to make it happen. If that happens too often, they'll be left waiting for customers who never seem to come.

Q-TIPS

Insurance companies should develop policies and procedures for choosing vendors such as independent adjusting firms and law firms. They should be extremely careful in selecting them, and the decision shouldn't be made on price alone.

The world-class organization has a formal supplier certification process in place. Suppliers should be fully aware of the organization's quality and service requirements.

The carrier should get assurances in writing from the supplier as to response time. In remote areas, it is sometimes better to rely on an established vendor who is farther away but can be counted on to provide excellent service.

Supplier quality assurance programs are based upon three key elements: quality, cost, and schedule. You don't select a vendor solely on the basis of price. You're looking for the best balance of those three elements. If vendors aren't able to meet the performance you expect in dealing with customers, you'd better find others who can.

Suppliers should be provided with feedback regarding their performance. There should be a formal system in place to provide that feedback.

CHAPTER 28

Customer Expectations

At a wedding, a slightly tipsy guest goes up to the groom with an envelope. "I hear you're going to Bermuda for your honeymoon," he says, slurring his words. "We want you to have a great time...a really great time."

The groom smiles uncomfortably as the guest continues the buildup. "You take this and spend it in Bermuda. Buy yourselves a great dinner or jewelry, but have a great time."

The groom accepts the gift graciously and puts it with the others. A week later, when he and his new bride open it, the man remembers the speech. They open the envelope with anticipation and pull out a $20 bill.

The couple is surprised at the size of the gift. They wouldn't have given it a second thought had the guest at their wedding not made such a fuss over giving it. He acted as if he was Ed McMahon, bringing them a check from Publisher's Clearinghouse.

Expectations have a significant effect on someone's reaction. The same phenomenon holds true when it comes to customer service.

At an expensive restaurant, you have higher expectations. You expect the quality of the food and service to be far superior to a restaurant where the prices are modest.

There's a poster that says: "The bitterness of poor quality lingers long after the sweetness of a cheap price is forgotten." And we've all felt that from time to time.

You buy a cheap piece of furniture, and it doesn't last. Within a year, the fabric wears thin or tears. Or perhaps that bargain cassette player eats the tape you're playing. You open up the lid and find yards of tape unraveled in the machine.

If you think it's bad that the cheap item is of poor quality, it's worse when you pay top dollar for more of the same. You pay the extra bucks for better tires or a vacuum cleaner, and the quality isn't all that great.

There's a rapidly growing fast-food franchise that seems to be opening up on every corner. Yet absolutely everyone we've talked to has made the same comment: "The food's good, but it's awfully expensive for chicken."

Or you go into an expensive department store and are ignored by the staff. Quality and price do not always go hand in hand. One of the companies with the best price on auto insurance in Pennsylvania also has a reputation for the best service. Furthermore, a company like USAA is well known for having low premiums and exceptional service. According to the *Harvard Business Review*, USAA has developed a reputation for superb service in an industry known for its indifference to the customer.

Customer expectations are greater when the product costs more. You don't expect gourmet food when you're paying fast-food prices. Nonetheless, at certain fast-food franchises, you expect a certain level of quality and service, no matter which location you choose.

No one expects bad service when buying insurance, no matter how inexpensive it is. Customers expect reliability and dependability. You file a claim, and it's resolved quickly and fairly. Your statements arrive on schedule. They're easy to read and accurate.

Customers expect responsiveness. Questions are answered in a timely manner. Phone calls are returned in a reasonable amount of time.

Customers expect competence. Some customers know absolutely nothing about insurance. They shouldn't be at the mercy of insurance personnel who know next to nothing, while the customer pays for their mistakes.

A company can intentionally build up your expectations of service. In a brochure for its MasterCard and Visa, a bank promises unparalleled customer service. Customers will get "Platinum Plus service."

The brochure spells out what that means. An exclusive toll-free number is available 365 days per year at all hours of the day or night. Though it's not spelled out, we assume they're available all 366 days in a leap year. If you're out of the country, you can call collect.

The brochure also promises an accurate statement that is easy to understand. We wonder if the bank is admitting that its lesser credit cards don't provide accurate and easy-to-understand statements.

In all seriousness, the bank may deliver exceptional service that lives up to the hype in the brochure. Our only concern is that companies live up to the higher level of expectations that they create.

Over the years, insurance companies have promoted deluxe policies geared to the affluent homeowner. The fanfare for these policies also builds expectations that service will be friendlier, faster, and fairer.

At some insurance companies, the credo seems to be, "Don't expect too much. Don't expect too little."

These companies meet their customers' expectations, more or less. On a good day, service is a pleasant surprise. On a bad day, they frustrate customers with their handling of routine matters.

It isn't hard to identify customer expectations. You don't need sophisticated market research. Just ask your employees, and they'll tell you.

Employees can give you hundreds of customer expectations, if you'll sit down with them. They may not always meet those expectations on a day-to-day basis, but employees know what customers want.

Customers want low prices but good service. They want to understand the bills from the insurance company, as well as the application for coverage. Customers want to deal with polite and knowledgeable employees.

The vast majority of customers are willing to accept that mistakes do occur. They don't, however, want to spend a lifetime, or sometimes longer, trying to correct those mistakes. The insurance company shouldn't make it torture to correct a mistake, but some do.

As part of her master's thesis, Beth Chung asked employees of a property-casualty company to list their expectations. As one might expect, Chung found that customers want empathy. They don't want to be put on hold more than once.

Employees have a wealth of knowledge about what customers want. Although employees may not always give customers what they want, employees know what it takes to provide great service.

Insurance companies also know what it takes to provide great service. Unfortunately, the commitment to service at some companies is like a pendulum that swings back and forth. At certain times, the carrier will place an emphasis on service and will limit the number of files that employees handle. When the pendulum swings toward controlling costs, employees are overwhelmed with many more files than the suggested workload.

There's a strong possibility that service will get worse as work is piled on those employees. Ideally, they'll be able to efficiently handle the additional workload, but the carrier is contradicting itself. The company has told employees how many files they should be handling and is now saying that all bets are off. In essence, the carrier has stated its expectations of employees and is now changing the rules.

Q-TIPS

Suggested workloads shouldn't be altered arbitrarily. The employer and employee should discuss why workload guidelines are being increased. Employees are entitled to know how much additional work is expected and should be given the opportunity to share their concerns about the situation. If changes in work distribution are made, they should be scrutinized to determine whether they have an adverse impact on customer service.

Both the company and the employee need to define what customers want and how to give it to them. They reach this definition by talking to customers and finding out what their expectations are. Successful companies stay close to their customers and listen to what they're saying. An insurance company should never raise the expectations of customers unless it can meet or exceed them.

CHAPTER 29

Expectations of Bad Service

There's more to the sitcom "Seinfeld" than just laughs. You can learn a great deal about customer service.

Perhaps the New York experience illustrated in "Seinfeld" isn't a true reflection of customer service in Manhattan. But for a moment, let's look at the negative examples offered by the hit comedy series.

The "soup Nazi," as Jerry labeled him, didn't quite grasp the concept that the customer is always right. Customers lined up for the establishment's soup, despite the owner's penchant for throwing people out who didn't please him. If you dare to question him for not giving you bread with your soup, you could be asked to leave.

The soup Nazi didn't have the hang of customer service. You followed his rules for ordering or bought your soup somewhere else. If a customer wasn't to his liking, he dismissed the customer from his establishment.

Despite his behavior, the customers came in droves and weren't driven away by his lack of courtesy. After all, his soup was to die for or at least worth giving up your girlfriend.

On another episode, the cast of "Seinfeld" walked on eggshells with the merchant who sold the best fruit in New York. He arbitrarily chose which customers were barred from buying his fruit.

And in yet another episode, Jerry and the gang waited to be seated at a Chinese restaurant for hours. The person taking names broke promise after promise to seat them shortly.

In real life, businesses depend on customers, not vice versa. In a sitcom, they can treat customers with disdain, but that's rarely the case with companies which do business in the real world.

Ironically, the "Seinfeld" cast was drawn to businesses with bad service in a town with thousands of restaurants and where one would think competition is great. Rarely will your customers be so fixated on your product that they'll overlook how you treat them.

It's hard to imagine any insurance company being able to treat customers with disdain. There are thousands of insurers out there who are licensed in the state to do business. The recipe for good customer service is consistency and fair treatment of everyone who deals with your company. Chances are, you know already which employees have the best customer-service skills. And you also know which ones bring their problems from home to work or take them out on the customers. It's time for you to resolve those issues before more of your customers deal with your own version of the soup Nazi.

Start with that health claims processor who's not a morning person. Even after three cups of coffee, she's still snapping at customers who had the audacity to call before ten in the morning. As the day goes on, however, she's patient and helpful with customers.

Or ask this question: Who's on first? No, it's not an Abbott and Costello routine.

At one insurance agency, it's a receptionist who picks up the phone. She was hired because she makes a nice appearance, except there's one problem. Eighty percent of her customer contact takes place on the phone.

This receptionist is the one who's on first with potential clients. What they hear first is the voice of someone who answers over 300 calls in a day. The receptionist rattles off the name of the firm and sees how quickly she can pass callers off to a customer-service representative.

Only it's the customers who feel they're getting short shrift. Their first exposure to the agency is an employee who can't wait to unload them on someone else. And the receptionist is even more abrupt when she has a personal call on the other line or has a lot of other work to do besides answering the phone.

At a large insurance company, it may not be a human being who's on first with the customer. One insurer refers to the automated response as the "easy dial" system, which may be one of the greatest misnomers of all time. Frustrated callers give up after hearing so many options; they forgot why they called.

Every insurance company has employees who are notorious for providing bad service. It's no surprise when a complaint about that person winds up on the supervisor's desk. Whereas there are always people who are better at their jobs than others, no organization should tolerate people from whom you expect bad service.

Q-TIPS

Expectations of bad service are no laughing matter. When customers have come to expect bad service from your company, they'll take their business elsewhere when the policy comes up for renewal.

Good customer service shouldn't be dependent upon the employee's mood or the time of day. Customers shouldn't be forced to walk on eggshells with certain members of your staff.

CHAPTER 30

Fair and Equal Treatment of All Customers

Consistency and fair treatment of customers start at the top. In Florida, Nationwide is accused of denying coverage to consumers because they are old, divorced, disabled, or live in undesirable areas. It not only violates the state's antidiscrimination laws, but a business practice like this would be inconsistent with the principles of Total Quality.

With quality management, you're trying to improve relations with customers, not alienate them. While every business has a right to weed out undesirable customers, it cannot and should not discriminate. According to agents who spoke with state regulators in Florida, Nationwide had a point system that indirectly discriminated against certain groups.

In some states, insurers refuse to write certain individuals because they have a bad credit rating. The person with a bad credit rating may find it difficult to buy certain policies.

The consumers who are turned down for a policy will almost always believe they are being treated unfairly. They'll protest that their driving record is spotless and that bad credit has nothing to do with insurance. Some will even offer to pay for the entire year in advance.

These people walk away believing they've been treated unfairly. They believe that they're being discriminated against because of their credit rating. Whether they have or not, their perception is that an injustice has been done.

In every transaction, all of us want to and need to be treated fairly. We're less than thrilled when we pick the wrong line at the bank and wait ten minutes longer than someone else who came in after us. Customers want a system that ensures fair and equitable treatment of customers.

Part of customer service is not offending the individual's sense of fairness. Businesses owe an explanation to customers who perceive they have been treated unfairly.

The only requirement in some states is that a one-page cancellation notice be sent to the customer. In two sentences, the customer will be told why he is now a former customer. He is also told of his appeal rights to the insurance regulators in that state.

The refusal to write a policy will be equally brief. The person with a bad credit rating will rarely be told why the credit rating has an impact on whether the insurance company will write the business or not.

There may be a valid reason why someone with a bad credit rating is a poor risk. The company may have empirical evidence that people with poor credit are more likely to submit claims or drive carelessly. If there are legitimate reasons for refusing to write this type of business, and insurance regulators permit this type of business practice, the insurance company can choose to make an underwriting decision of this kind.

Nevertheless, the policy should be applied across the board and not used selectively or in a discriminatory manner. Furthermore, the customer deserves a better explanation than two sentences on a refusal to write notice.

If it's not provided, customers will believe they've been treated unfairly. And whether they're former customers or people who expressed an interest in doing business with your firm, they deserve better treatment than that.

Customer service doesn't mean accepting every customer. Insurance companies have always been selective in deciding which risks to insure. The underwriting process is based on the premise that you aren't going to insure everyone. There must, however, be a fair and rational process that doesn't violate the applicable insurance regulations.

There are some who believe that insurance companies actually operate with a hidden agenda. They justify their actions using legitimate reasons, but their underlying agenda is more sinister. For example, an insurer may be engaged in redlining but manages to cover up this practice.

As another example, in most states an auto policy may only be nonrenewed for specified reasons, such as two at-fault accidents in 36 months. A company that wants to nonrenew a particular insured may not have a basis to do so under state law, so it takes another course of action. The insurer sends questionnaires more frequently to the insured, hoping that the person won't return one, which opens the door to cancelling the policy for lack of cooperation.

An insurer that wants to get rid of a particular insured may wait for the moment when the individual is a day or two late on his premium payment. Whereas the insurance company would forgive most policyholders who are a day or two late, it has been waiting gleefully for this insured to miss the deadline. The carrier cancels for nonpayment. In many states, there is no grace period when an auto insurance premium is late.

Customer service isn't about lying in wait for the policyholders who aren't as desirable as the company once believed. It's about fair and consistent treatment. The insurer follows the letter of the law, as well as the spirit. Companies don't exploit technicalities to weed out customers.

Q-TIPS

Although insurance regulations may permit certain business practices, an insurer should examine each of its policies and procedures to determine if groups or individuals are treated unfairly. An underwriting practice may have outlived its usefulness.

Even if the insurer must drop a customer, the individual deserves an explanation of the reasons for that decision. Complying with the state's rules regarding cancellation, nonrenewal, and refusal to write is just the minimum level of customer service.

Chapter 31

Fair Warnings and Quality

Traditionally, quality service adds a personal touch. Many customers get that personal touch from their agent, but how about from the CEO of the insurance company?

One insurance company chairman took matters into his own hands, according to *The Wall Street Journal* on April 17, 1996. Robert S. Weiss, the CEO of Connecticut Life & Casualty Insurance Company, met with young insureds involved in accidents. Weiss then proceeded to scold them and warn them about the dangers of unsafe driving.

The *Journal* called it a form of tough love in the insurance industry. Weiss also met with young insureds cited for traffic violations and even a few newly licensed drivers. Weiss warned the young drivers that they might be nonrenewed and that their parents' rates would go up.

Whereas some would disagree with Weiss's approach, there is one thing that's clear. The insureds were warned about the ramifications of their behavior. In the last chapter, we discussed that fairness is an important component of quality and customer service. Fairness also involves warning customers that their coverage is in jeopardy or that their rates may be escalated.

In too many instances, insureds aren't given a warning, and that's bad service. They're not told that if they submit another claim, even a small one, they'll be nonrenewed. No one tells them that their policy will be cancelled if they don't return the questionnaire.

Important points are overlooked in the sales process too. The agent never advises the new customer that the company can refuse to write the policy in the first 60 days if it doesn't like the individual's credit rating or she has an accident. By the time the insured finds out that the company is refusing to write the coverage, she has already cancelled her old policy.

Similarly, some agents selling annuities never point out the surrender charges or that the attractive rate is only guaranteed for one year. If the customer doesn't ask, pertinent information isn't mentioned. Whereas insurance regulations may require disclosure of pertinent information, the agent may thrust a stack of documents in front of the client for signature. As a result, the pertinent information is conveniently overlooked.

The agent selling long-term-care insurance stresses that the premium is based upon the current age of the customer and that it won't go up with age. The agent neglects to mention that the premium can be increased if all policyholders in the same classification receive a rate hike.

We're not talking about an overt misrepresentation that usually occurs in a churning situation. The sins of omission are also a major concern. These sins of omission, such as failing to mention important facts about the policy, undermine quality in the selling process.

Agents will protest that no one will ever buy a policy if they point out every negative feature. You don't sell insurance by accentuating the negative. For example, not many people will switch auto policies to save a few dollars if they know the carrier has 60 days to evaluate whether it wants to insure them. In some states, even though the agent has written the policy, the carrier can still refuse to continue the coverage for up to 60 days.

The quality agent or insurance company warns customers of potential problems. They warn auto insurance policyholders that a late payment, even of a few days, can result in cancellation.

Similarly, quality insurance companies keep insureds advised of potential problems. They send warning letters, advising insureds that a lawsuit may exceed the policy limits. Obviously, this is the right thing to do from a quality perspective. Furthermore, the insurance company might be on the hook for a verdict in excess of the policy limits if it doesn't disclose this information to the insured in a timely manner.

The insurance policy gives the carrier discretion to settle claims and lawsuits against an insured. Nevertheless, thousands of insureds are surprised to learn years later that the company paid out money on certain claims.

In some instances, the other driver was reimbursed for damage to his bumper or even received thousands of dollars for an injury that the insured did not necessarily cause. The insured never even knew that the third party was paid and that this might have an adverse effect on the insured's record with the company. Many policyholders never find this out until receiving a cancellation notice.

There should be few surprises when it comes to quality. While customers may not be totally satisfied with the outcome, that outcome should not come as a complete surprise.

Q-TIPS

Along with the disclosures that are required by statute, agents should create their own checklist for each policy they sell. They should keep a copy in each file, indicating that the client was warned about the potential negative features of a particular policy.

Companies should keep policyholders informed about pending litigation. If a third-party claim is settled, the insured should also be advised.

CHAPTER 32

Consistent Coverage Decisions

If you ever want to get the conversation rolling at a party, mention health insurance. Everyone will have a horror story about bills that weren't paid but should have been paid. Worse yet, the so-called explanation of benefits sometimes leaves the most intelligent people unsure if the doctor was paid.

With health insurance, one wonders how many errors occur every day. The state passes a law which mandates that health insurers pay for a woman's routine gynecological visit. If the gynecologist discovers a problem during that visit, the health insurer now says it's no longer a routine gynecological visit and doesn't want to pay the physician. After months of improperly processing hundreds of bills, the insurer agrees that it should have been paying for these initial examinations and blames the mistake on a coding error.

Small medical bills can cause chaos. They are passed from the health insurer to the workers' compensation carrier to the patient's auto insurance. Sometimes claims processors at the health insurer just don't understand the employer's particular insurance plan.

The insurance industry knows there's a problem. One insurer has dozens of employees working in its recovery unit. They routinely attempt to collect overpayments. To date, the insurer hasn't felt compelled to create an underpayment unit.

Across the country, businesses exist to help people with their medical insurance problems. They offer assistance to the thousands of people who can't make sense of their health insurance and need help. As is the case with many forms of insurance, even bright people feel they are in over their head.

Health insurers blame these mistakes on human error or faulty computer programs. Maybe it's just a deductible or co-payment that's misapplied. Or maybe it's a life-or-death mistake.

On February 17, 1994, the *New England Journal of Medicine* reported that health insurance companies make inconsistent decisions about coverage that benefits the insurer rather than the insured. Of 533 breast cancer patients seeking approval for bone marrow transplantation, 121 were denied on the grounds that the procedure was experimental. The decisions on coverage were, according to the study, inconsistent with the clinical characteristics of the patients and the type of treatment they were receiving.

When it comes to managed care and utilization review, poor quality can result in death. Although managed care seems to be in everyone's best interests, there are always going to be mistakes. A utilization review nurse at the insurance company denies a request for a particular kind of treatment, and someone dies. The defense is that it's not the insurance company that's responsible for the death, because the individual can still go for treatment. The reality, however, is that many people won't seek treatment if they can't afford the cost. Furthermore, some people will rely on the nurse's decision to their detriment.

Employees at one California HMO have a dark sense of humor. After several bomb threats, they joked that the calls were made by the relatives of HMO members who died while waiting for treatment to be approved.

Quality errors in the insurance industry are a deadly serious issue. They can lead to bankruptcy and human heartache of immense proportion. Sometimes people lose their lives or their life savings.

What's the solution? Sometimes the answer is legislation. When insurers were pushing mothers with newborns out of the hospital too quickly, a legislative solution was sought.

Better training for employees who process claims is a good start. And those trainees shouldn't be let loose on the public until they've passed an exam. Their work should be reviewed frequently until it's clear they've got the hang of how health insurance works.

Perhaps all those job ads looking for applicants should ask for three to five years of experience, not one year. Whereas a casualty-property claim representative position might require a college degree, applicants for the claim processor job normally only need to have a high school diploma.

Better software will help too, and not just to weed out the unintentional errors. A growing number of insurers utilize software to weed out fraudulent and questionable claims. According to some studies, health insurance fraud costs the industry over $50 billion per year.

There should also be explanations of benefits that are in plain English, so the insured can understand what action has been taken. The insured will be able to understand if the action is appropriate.

Insurance companies should offer a meaningful mechanism for disputing a decision. The grievance procedure shouldn't simply be a rubber stamp of the claims-processor's decision.

If you ever want to bore people to death, make them listen to two insurance people argue about a complex coverage question. The argument will put most people to sleep in moments.

Most people don't realize that issues in insurance policies aren't black and white. Some of the contract language can be interpreted in different ways. You'll find seasoned veterans of the insurance industry with totally different views on how the policy should be interpreted.

Some insurance companies interpret their policies narrowly. Others give their customers the benefit of the doubt when the policy is ambiguous on a particular point. A few companies look at the size of the claim and decide whether to give their customer the benefit of the doubt.

If the coverage question goes to court, ambiguity is to be resolved in favor of the insured. The insurance company wrote the contract, so if the wording is unclear, that's not the insured's fault.

Obviously, even consistency has to give way to real-life problems. If you're dealing with an inflated or questionable claim, this insured shouldn't be given the benefit of the doubt.

As we've mentioned earlier, customer satisfaction is often a product of the result. A favorable result means that the insured believes the company did a good job.

Coverage denials make insureds believe they paid money for no apparent benefit. Insurance is intangible to begin with, and coverage denials make its value seem totally nonexistent.

Although you'll rarely be able to deny coverage and keep a happy customer, you need consistency to maintain quality. Coverage decisions should not vary from supervisor to supervisor or from claim representative to claim representative. The department, as well as the company, should be making consistent decisions on coverage, unless there is a court ruling peculiar to a particular jurisdiction.

Q-TIPS

Many insurers are extremely bad at communicating coverage decisions to everyone in the department. There should be ongoing meetings where coverage questions are discussed. Employees should be encouraged to bring coverage questions to the meeting. If there is no definitive answer, the issue should be resolved in conjunction with the legal staff, and the outcome should be shared with everyone in the department.

No organization should rely on the grapevine to circulate important news about court decisions or new legislation. A mechanism should be developed for distributing information to members of the staff, whether it's a newsletter or weekly update. The document might also include common problems that have been uncovered during quality reviews.

CHAPTER 33

Customer Feedback and Complaint Handling

The complaint letter comes in, or at least it looks like a complaint letter. It's handwritten and nine pages long. In painful detail, the letter recounts every purported error made by the insurance company in handling the account.

The customer operates under the impression that providing every detail enhances the credibility of the story. It reads like an episode of "Dragnet" with Jack Webb giving the date and time when each event occurred.

By the end of the letter, the reader is not quite sure what the complaint is, yet must respond to the complaint. Somewhere, buried under the minutia, is a complaint about how the insurance company did wrong by this customer.

At least this customer's letter got to the right place. Another angry customer writes a letter of complaint to the president of the company. It never gets to that individual or even a member of her staff. At some companies, the complaint letter actually works its way back to the person who handled the file, the individual who quite possibly is part of the problem.

At another company, the president's secretary passes the complaint on to a subordinate. There's no control number on the document or any way to verify that the subordinate followed up with the customer. No one is required to send a copy of the response to the president.

In some organizations, customer feedback and complaints warrant a letter of explanation or apology. Management feels it's the right thing to do, so Kato prepares the response:

Dear Customer:

We're sorry you feel this way. Bummer!

Sincerely yours,

Kato

Well, the letters aren't usually that bad. Some are worse. The same form letter is sent in response to every complaint and doesn't give the customer any hope that the problem is being addressed.

A good start is to let Kato and other employees take another stab at personalizing these letters. Here's Kato's next attempt:

Dear Customer:

We're sorry if we "dissed" you with our handling of

your account. Sometimes we can be pretty lame.

Sincerely yours,

Kato

No, that's not quite what we have in mind, yet businesses don't do much better. The problems start when a husband and wife arrange for a new certificate of deposit. With the new CD, they are to get a number of free services from the bank. When they are charged for those services, the couple tries without success to resolve the problems through the bank's 800 number and through the branch manager.

After having no luck, their last resort is an employee who functions as the bank's ombudsperson. He finally resolves their problems in an exemplary manner.

The customers write a letter to the bank's president describing the problems and how they finally were resolved. The bank's response to their letter deals exclusively with the two paragraphs that praised the employee. The bank thanks them for bringing the employee's performance to its attention, while ignoring the four other paragraphs that addressed customer-service problems. The letter ends with, "Customers like you are a pleasure! We appreciate your business and look forward to serving you again."

Instead of getting misty-eyed over the display of affection, the customers shake their heads and wonder why the other issues they raised weren't addressed. They never would have needed the help of the employee they praised if all was right in the land of customer service.

Insurance companies make many of the same mistakes. There's no particular person designated to deal with complaints. Instead, customers wishing to register a complaint are frequently passed from one unit to the next. Unless the customer knows the inner workings of the insurance company, his complaint only makes it as far as the person he dealt with or that person's immediate supervisor.

In theory, answering complaint letters is easy. The response should be sensitive to the customer's position and must demonstrate empathy. It should answer all of the questions and issues raised.

The response should look professional. It should be grammatically correct. The spelling should be perfect, especially when it comes to the name of the person making the complaint. Sentences should be short and to the point, taking care to avoid insurance jargon.

Responses should be timely. If you need to investigate, acknowledge the complaint and assure the person that a response will be forthcoming in a specified number of days.

Admit when errors were made, and apologize. Let the customer know how to reach you if there are still questions.

Even if you follow all the rules, you're not going to satisfy every customer. No matter how logical or justified your position is, some will still be unhappy.

Some complaint letters are scary. The person writing them is either unable to communicate or may be a little off psychologically. These complaint letters make veiled threats against the company or the individual who is perceived as the problem.

Few insurance companies use complaints as a training tool. Oh sure, someone may pull them out at a subordinate's performance appraisal, but they are rarely used constructively. Companies that want to improve quality on a shoestring budget might start by collecting complaint letters and using them as a springboard for discussion.

Ironically, a few companies will spend thousands of dollars on surveys but will ignore customer complaints. Complaints are an important tool for gauging customer satisfaction.

The vast majority of dissatisfied customers never complain directly to the company about service. Instead, they simply take their business elsewhere and advise their friends to do likewise.

There is also a tendency for companies to ignore ex-customers. When insureds switch carriers or close accounts, no one wishes them well or bothers to ask why they're taking their business elsewhere. The opinions of these former customers can provide volumes of information about whether your company met their expectations. Furthermore, there may still be time to recover that lost business.

There should also be training on how to respond to complaints. Once again, you're not providing a stack of form letters for personnel to use, but guidance on how to defuse anger and resolve problems.

There should be policies and procedures dealing with how to respond to complaints. The employee who responds to a complaint should be empowered to cut red tape or to help resolve an ongoing problem where the customer has a legitimate beef.

Although no organization likes to get complaints, they do represent at least one customer's perspective, albeit an angry one. While complaint letters aren't necessarily representative of the typical customer's experience, they can't be dismissed by saying that only the angry people write and the happy ones don't.

While you don't want to invite complaints, an organization should make it easy for customers to voice concerns. A system should be established so that complaints of every kind can be channeled through one particular phone number or post office box. The system should ensure that someone in authority, not the same people involved in the situation, will be reviewing the matter.

Perhaps each division can establish a particular unit to review complaints and look for patterns. The people in that unit will attempt to resolve problems. They can follow up with the customer after the problem has been resolved to see if all of the issues were addressed.

As mentioned earlier, the American Council of Life Insurance has developed Principles of Ethical Market Conduct and an accompanying Code of Life Insurance Ethical Market Conduct. Although the Principles and Code only apply to individually sold life and annuity products, the concepts contained therein might be used by any insurer, no matter what type of insurance it sells.

There are six principles that must be adhered to by life insurance companies that subscribe to the Principles and Code. **Principle 5** requires the company:

> *To provide for fair and expeditious handling of customer complaints and disputes.*

Principle 5 obligates the insurer to develop policies and procedures for resolving complaints about market conduct. The insurance company must provide a mechanism for handling complaints and disputes fairly, as well as in a timely manner.

Under Principle 5, complaints must be identified, evaluated, and handled in compliance with the applicable state laws and regulations. Most states have implemented statutes that protect against unfair insurance practices. These laws penalize insurers who don't act accordingly.

To abide by Principle 5, the insurer must make a good faith effort to resolve complaints and disputes. The insurance company can't force the individual to file a lawsuit without first trying to resolve the situation amicably.

The logic behind Principle 5 is sound. A large insurance company has many lawyers on the payroll. The insurer has the resources to contest a case in court for years, if it so desires. It would be a violation of Principle 5 to do so, unless the insurer has attempted to resolve the matter first.

Under Principle 5, the company must make it easy for customers to complain. There must be a complaint process in place, and that process must be communicated to customers. There should also be a process to solicit customer feedback. The following are examples of useful tools to satisfy Principle 5:

- Use of an 800 number
- Mailings with pre-addressed return envelopes
- Customer response cards
- Accessibility beyond normal business hours
- Customer contact that coincides with policy events
- Customer interviews and focus groups
- Surveys following the sale

There is an assessment process to determine if an insurer has complied with Principle 5, as well as with the other requirements. During the assessment process, the assessor will look for objective data that proves the insurer makes it easy for customers to complain and that it solicits feedback.

Principle 6 obligates the insurer to do the following:

> *To maintain a system of supervision and review that is reasonably designed to achieve compliance with these Principles of Ethical Market Conduct.*

The insurance company must set up a system to supervise the market activities of distributors and employees involved in the sales process. If these people are not complying with the Principles and Code, the insurer can't just throw up its hands and claim that it was unaware of these violations. The insurance company must have policies and procedures in place to make sure this doesn't happen.

There should be policies and procedures in place for monitoring sales practices of employees and distributors. This may be accomplished by conducting random interviews with customers or follow-up surveys, by identifying trends, or by some other means. The insurance company is required to track customer complaints involving a distributor.

The aggressive steps taken by the ACLI to improve market conduct and compliance will help any insurer improve customer service as well.

Q-TIPS

Putting control numbers on all complaints will help management track responses and be certain that contact was made. Form letters should be avoided whenever possible.

All complaint letters, as well as surveys, should be read thoroughly, taking care to highlight all issues raised so they can be addressed in the response. The quality review process should include an evaluation of the responses to customer complaints.

Analyze what you learned from the complaint, if anything. Make changes in the system, if warranted.

In a world-class organization, information gathered from customers is analyzed and applied to its fullest extent. There is a formal process in place to improve relationships with customers.

CHAPTER 34

Quality Initiatives

At some companies, the executives think they know the answer to improving quality and customer service. The walls are filled with posters promoting quality, teamwork, perseverance, and leadership. They hand out notepads and coffee cups with quality messages on them.

There's even a fanny pack with a quality message on it. Under their breath, employees mumble that the quality program belongs in the same general area as the fanny pack.

Everyone wears lapel pins, affirming the company's commitment to excellence and quality. One vice president even has a rock on her desk with a quality slogan etched into the design.

To some, the pins symbolize something other than quality. They've become part of the corporate uniform for the movers and shakers who mimic the way their bosses dress. Instead of being viewed as a commitment to quality, service, or whatever, the pins are perceived as a symbol of blind obedience to whatever ill-advised policy is formulated.

At all-employee meetings, the company shows motivational films featuring a successful football coach. At another, a retired baseball star talks about the ingredients of a winning team and applies those same principles to this particular insurance company.

After the meeting, many of the female employees shake their heads in disbelief. They wonder who chose a speaker that few women employees could relate to and feel snubbed by the continued use of sports metaphors in every speech.

At the same time, a business magazine does an interview with the chief executive in his beautifully appointed office with expensive works of art. The CEO promises continued downsizing, much to Wall Street's glee. The article is passed from employee to employee, each of whom wonders if he or she will be affected by the latest round of cuts.

Teamwork is the one of the keys to quality, but it's not achieved with hollow tokens. At many companies, the team keeps getting smaller, and the players change from day to day. Employees find themselves playing their own position, as well as a former teammate's.

Certainly, there is a place for medallions to recognize employees or knickknacks to remind everyone of the company's vision. Nevertheless, these ornaments won't inspire teamwork if management sends dozens of other signs that the corporation consists of the haves and the have-nots.

At one company, the rumor of layoffs is in the air again. Productivity suffers once more, and finally, the axe falls. Once again, a few more of the worker bees are affected, and the executive ranks remain unscathed. Once again, there seems to be no rhyme or reason to the cuts.

In a *Dilbert* cartoon, Scott Adams pokes fun at symbols of quality. A character puts a coffee cup down on a piece of paper and creates the company's new logo: the brown ring of quality.

At too many companies, quality initiatives are symbolic and have very little substance. Yet company leaders expect employees to rally around the brown ring of quality or whatever meaningless symbol is the focal point of the effort.

There was a popular movie in the 1970s called *The Stepford Wives*. In the film, the men in a small suburban town grow weary of their wives when the women begin thinking for themselves and want to be independent. They long for the "good old days" when women obeyed their husbands and worshipped them.

In this sci-fi film, these shallow men have a solution to their so-called problem. They are able to replace their wives with duplicates who obediently serve their husbands, wishing only to please them sexually and keep the house clean.

The Stepford wives look almost exactly the same, except for a few enhancements that the men prefer. Eventually, even Katherine Ross, the star of the film, is turned into a Stepford wife who is incapable of thinking independently.

In many companies, the executives are looking for Stepford employees. They promote the employee who hangs on their every word and never questions any policy, no matter how ludicrous. Independent thought and creativity are discouraged.

You can benchmark until you're blue in the face, but innovation is still indispensable. Whether you're brainstorming solutions to a quality problem or looking for ways to improve service, creativity and ingenuity, not blind obedience, work best.

Quality improvement demands fresh approaches, not the same tired old ways of approaching a problem. Employees shouldn't be pressed for solutions that will never be implemented. You'll create a generation of Stepford employees if you make them spin their wheels on problem-solving exercises that are meaningless.

Q-TIPS

Cynical employees will be resistant to Total Quality initiatives. Cynical employees are made, not born. The leaders of the organization may unknowingly be fostering an atmosphere of cynicism.

It is important to bring that undercurrent of cynicism to the surface. At individual and group meetings, managers should elicit feedback on the quality initiative and must try to find out why employees are skeptical. Although some employees will always be negative, you can bring those feelings to the surface rather than allowing them to fester.

CHAPTER 35

Suggestion Boxes

You can't just put out a suggestion box and say you're improving quality. Inviting employee input to improve quality is an important step, but there's more to it than that.

In 1992, Zurich-American introduced its "What's Z Big Idea?" campaign. The employee idea campaign featured a cartoon dog named Spot. Spot, Z-idea dog, was the focal point of a program to generate suggestions for improving quality.

The insurer gave out Spot awards for suggestions. Employees submitting ideas selected for implementation received additional prizes.

As part of every quality initiative, it seems there's a plan to encourage employee suggestions. Companies should be applauded for attempting clever approaches and not just leaving a suggestion box in the lobby.

The danger, however, is that employees will be frustrated if suggestions aren't implemented. At one company, a committee reviews suggestions to see who wins the $10 cash award. Suggestions viewed as frivolous are tossed out by a human resources representative before the committee meets. After the committee selects a winner, however, the suggestion is passed along to management and is rarely implemented.

At that same company, good ideas aren't communicated between co-workers. They're held back by employees, hoping to win a few bucks for the best suggestion. When employees make suggestions that are rejected by the committee, they abandon the ideas, even though they could be used to make improvements in their area.

On many occasions, the individuals making suggestions will not always have the best intentions. They may have an ulterior motive other than just winning a prize for making a suggestion. Perhaps the suggestion won't improve customer service but will simply make their lives easier.

Although suggestion plans can undoubtedly help improve quality, it's quite likely they'll be frisky at first and dog-tired as they wind down. While a new campaign can sharpen the focus on quality, you might need a new prescription after a few months.

Improvement needs to be continuous and ongoing. It requires extensive employee involvement. Zurich-American even had a continuous improvement division working toward improving quality.

Although new wrinkles must be added from time to time, it should be clear that the quality improvement process is ongoing. To avoid the program-of-the-month mentality, it should be made abundantly clear that the old dog still can hunt. You're simply trying to remind people about quality with these new programs. You can teach an old dog new tricks about quality and customer service.

Q-TIPS

Suggestion boxes are useful but sometimes outlive their usefulness. It's dangerous to reward individual suggestions. Encourage team suggestions and reward the group.

Suggestions should not be arbitrarily discarded by someone who may not recognize their value. Although every suggestion can't be implemented, it is important for employees to see that a number of them are used.

Chapter 36

Management Support

In our favorite beer commercial of all time, a man and his dad are fishing. In one of those moments that guys rarely have, the son wipes away a tear from his eye and says with as much sincerity as he can muster: "I love you, man." The father ignores him and advises his son that he can't have his beer. The outpouring of feigned affection just isn't enough to make him part with this particular brand of beer.

You sometimes see a comparable situation when the new quality plan is announced. Whether they believe in it or not, management always endorses the quality improvement initiative with real or feigned enthusiasm. Everyone realizes that management support is indispensable, but there's more to it than executives standing up at an all-employee meeting with tears in their eyes to say the equivalent of "I love you, man."

In some organizations, management's support for quality is every bit as insincere as our friend in the beer commercial. Even if management is enthusiastic at the beginning, that support sometimes erodes. As a result, the foundation of the quality initiative crumbles.

At a quality kickoff meeting, the president of the company stands up and talks about his undying support for Total Quality. He pledges that this is the dawn of a new era where quality is king. It's the last quality meeting he attends, however, leaving employees to wonder if quality wasn't king, but just "queen for a day."

At other companies, quality initiatives are viewed as the "program of the month." Management introduces each and every one but then fades into the woodwork. Some managers sign in at quality meetings and then quickly excuse themselves to take phone calls that may or may not exist.

At some insurance companies, the commitment to quality is made with symbolic gestures. The CEO decides it's not enough to just have a quality coordinator or quality facilitator. You need a quality manager. No, wait, let's really show how committed to quality we are. The CEO appoints a crony as director of quality.

While symbolic gestures can show management support for quality, they are certainly not sufficient. Management support must be visible and sincere.

Titles, whether someone is the vice president of quality or whatever, are only important as a means of giving clout to the person spearheading the initiative. Titles are one way to empower the person leading the quality process.

Symbolic support is better than nothing. The CFO at one company gives this speech to the new quality manager. "I won't bother you," he says. "But in six months, if I don't see any measurable improvement in quality, then I'll be in your face."

This same individual is the only member of management who doesn't attend the rollout meetings for the quality initiative, nor can he find time to participate in the makeup session. Not so surprisingly, his people don't lend their support to the quality improvement effort.

When quality never seems to improve in the organization, members of management shake their heads and ask what went wrong. They rarely point the finger at themselves. They expected the quality initiative to fail, and it did. It never dawns on them that this was a self-fulfilling prophecy.

It's enough to leave people who believe in quality crying in their beer.

Q-TIPS

All key members of management must support the quality effort. It's not enough to give token support to the quality improvement process.

Be wary of hidden agendas. There are some people who want to see the quality improvement effort fail for some unarticulated reason.

In a world-class organization, members of senior management are personally and visibly involved in creating a customer-focused environment.

Chapter 37

Limping the Talk

The financial newspaper is delivered in a yellow plastic bag to the customer's doorstep. Printed on the bag is "Service is our priority."

Two bags sit on the steps of the home for two weeks. The subscriber called to stop delivery when she went on vacation, but the message didn't get to the circulation department in time. The man who delivers the publication drives by the house each day and delivers newspapers to other homes on his route.

It never dawns on him that he should pick up the two newspapers that were delivered in error. The delivery man gives no thought to picking up the two bags and holding them for when delivery is scheduled to resume. Instead, the bags stay there so potential thieves will know that the owner of the house isn't at home.

It's scary to think what service would be like if it wasn't a priority. In reality, service is an afterthought, despite what the company claims is the case.

There are quality and customer-service slogans by the hundreds. But for some, the so-called commitments to customer service and quality are just words.

In management circles, there are tons of neat little phrases that managers are supposed to live by. For example, you can't escape the current "in" phrase, which is to "think outside the box."

Another management cliché is the advice that you do more than just "talk the talk. You have to walk the talk." But in some organizations, all they do is "limp the talk."

Walking the talk means that you pay more than lip service to quality. You make public statements in support of the quality program. Upper management must demonstrate that it backs the quality initiative.

Upper management has to walk the talk for the duration, not lose interest quickly. While organizations never make lifetime commitments, there must be assurances that the quality improvement initiatives will endure for a year or longer. It may take many months for the results of improvements to be demonstrated.

One way to walk the talk is by creating self-directed work teams. Some organizations refer to them as semi-autonomous work groups or self-managing teams.

Self-directed work teams consist of 6 to 15 experienced employees who will assess the processes of the organization and make improvements. They're different from quality circles or employee involvement initiatives. Unlike those programs, participation on a self-directed work team isn't another duty thrown onto an already filled plate.

These work teams must have full access to information. A member of senior management should be given responsibility for supporting the team and obtaining resources for it, as well as keeping it on course.

Upper management must give the go-ahead on projects. Nevertheless, the team should be empowered to make decisions. If you want your quality initiatives to be up and running, you have to do more than just limp the talk.

Q-TIPS

Giving lip service to quality and customer service is never enough. There has to be management commitment, through both words and actions.

CHAPTER 38

Barriers to Quality Improvement

The characters in the *Dilbert* comic strip ask this question when assigned to a quality initiative: "Is there any risk this will devour our productive hours, lower our morale, and have no impact on profitability?"

The tone of the question more than suggests that they believe working on a quality initiative will have a negative impact. Their suggested name for the quality initiative is "Qualicide."

Judging by the number of people who post his cartoons on their bulletin boards and buy his books, *Dilbert* creator Scott Adams has touched a nerve. A lot of people are cynical about quality initiatives.

In theory, being against quality and customer service is like being opposed to motherhood and apple pie. Unfortunately, at too many companies, quality has been shoved down employees' throats, leaving a very bitter taste in their mouths.

At many insurance companies, employees are cynical about more than just quality initiatives. They've become distrustful of the company they work for and the executives who run it. As companies become less paternalistic in an era of downsizing, employees will view every program with skepticism.

Whether it's quality improvement, quality management, or Total Quality Management (TQM), employee cynicism and distrust are tough to overcome. They aren't the only barriers an insurance company faces in making improvements.

David Letterman has his Top Ten List, but Robert J. Masters, Dean of the School of Business at Southeastern Oklahoma State University, has outdone him. Dean Masters has collected the top 15 barriers to TQM implementation. Dr. Masters conducted an extensive review of the literature and found 15 distinct barriers to TQM. Unlike Letterman's Top Ten List, overcoming these barriers is no laughing matter. The barriers are as follows:

1. Lack of management commitment
2. Inadequate knowledge of TQM
3. Inability to change organizational culture
4. Improper planning
5. Lack of continuous training and education
6. Inability to build a learning organization that provides for continuous improvement
7. Incompatible organizational structure and isolated individuals and resources
8. Insufficient resources
9. Inappropriate reward systems
10. Use of a prepackaged program or inappropriately adapting TQM to the organization
11. Ineffective measurement techniques and lack of access to data and results
12. Short-term focus or band-aid solutions
13. Inadequate attention to internal and external customers
14. Inappropriate conditions for implementing TQM
15. Inadequate use of empowerment and networks.

Dr. Masters concentrates on the eight barriers that plague organizations most frequently. Those are: (1) lack of management commitment, (2) inability to change organizational culture, (3) improper planning, (4) lack of continuous training and education, (5) incompatible organizational structure and isolated individuals and departments, (6) ineffective measurement techniques and lack of access to data and results, (7) inadequate attention to internal and external customers, and (8) inadequate use of empowerment and teamwork.

If one or more of these barriers exists in your organization, the result can be "qualicide," as Scott Adams put it. Nevertheless, before you call Dr. Kevorkian, it might pay to listen to Dr. Deming, who urged people to remove barriers to quality.

Q-TIPS

Although many barriers to continuous improvement exist, they can be overcome by an organization that is willing to surmount them.

Chapter 39

This Company Isn't What It Used to Be (And It Wasn't That Great Before)

Employee attitudes can always be a barrier to quality improvement. "This company isn't what it used to be," the employee frequently laments. "And it wasn't that great before."

Jack's been whining about the company for as long as anyone can remember or at least since man has walked upright. You could chart his complaints over the years. During every decade of his employment, he's complained and compared the situation to days gone by. At every juncture, the previous era seems like the Golden Age of Greece when compared to the way things are now.

The previous management team or corporate ownership was always the paragon of virtue when compared to the drek that's now running the company. Ironically, some employees recall that so-called golden age during which Jack ranted and raved about the people he now speaks of with such high regard.

Many of the young employees complain every bit as much or worse. Based on their months of experience, the company is being run badly. During their short tenure, they've developed an attitude that rivals Jack's, which has aged like bad wine.

It truly is amazing how they've built up so much ill will in so little time. Perhaps the company reminds them of a previous employer they hated, so it didn't take too long for there to be a cumulative effect.

These newer employees have far less reason than Jack to hang on. They have very little invested in the company, financially or emotionally.

At a different insurance company, whining is part of the culture. Its employees are viewed as chronic complainers, though no one can figure out why they're so unhappy.

Although the company's reputation is good nationally, the employees have trouble finding other jobs in that city. They are viewed as spoiled, as well as overpaid and underworked.

For the most part, the employees never look for work elsewhere. They just moan and groan about how bad it is to work for that company. Unlike Will Rogers, they've never met a CEO they didn't hate.

There seems to be a love–hate relationship at work here. The hate manifests itself more frequently, yet not many of the employees leave.

It's not just psycho-babble to say that work should be more than just a paycheck. You'll never achieve Total Quality if employees are there because they can't or won't go anywhere else.

Employees need to believe in what they're doing. They should feel they're part of a team.

There should be a system of rewards that goes beyond pay. Formal and informal recognition should be an integral part of the workplace.

Not everyone wants to grow. A few want to do the same job over and over. They'll resist change of any kind. But those who do want to develop new skills should be encouraged.

Fortunately, all but the worst attitudes can be overcome. While employee attitudes can be a barrier to quality improvement, they can be changed in most instances.

There are some simple role-playing exercises that can help employees see how negative they are. Without singling out the problem employees, let them take turns as manager at a training session. Switch places with the employee and see how he or she copes with a constant complainer. You just might get positive results.

Q-TIPS

Difficult employees should be coached, not criticized. In a team-oriented culture, they may realize that their negative attitudes are hurting everyone, not just the targets of their diatribes.

It is useful to try to isolate the source of the problem employee's complaints. If he or she has a specific complaint, try to resolve it. If employees have complaints that they can't or won't let go of, determine whether that makes them unwilling or unable to do their job. When attitude affects job performance, it should be dealt with immediately.

In a world-class organization, senior management is aware of employee morale issues. The members of management will constantly work on projects to improve morale. They will develop measures and indicators of employee satisfaction. Wherever possible, management will try to eliminate the root causes of morale problems.

CHAPTER 40

Recognition of Employees

At many companies, the only recognition you get is at the time of your performance appraisal. *Dilbert* creator Scott Adams captures how some employees view the process.

In Adams' cartoon, Alice is told that her performance meets expectations, and she gets a 2 percent raise. Alice protests that she worked 80-hour weeks, earned three patents that made millions for the company, and donated bone marrow twice to their biggest customer. Her boss's only response is that he noted the bone marrow situation as an attendance problem.

Obviously, we can't do justice to Adams' cartoon. Similarly, companies frequently don't do justice to the employees in their organization who deserve to be recognized.

The creator of *The Far Side*, Gary Larsen, carried the employee-of-the-month concept to its illogical extreme. In his cartoon, Bob is named "torturer of the month" for his fine work in inflicting punishment.

At some companies, selecting the employee-of-the-month is pure torture. The members of the general manager's staff are meeting to pick the employee-of-the-month. They've got specific criteria to choose from in making the selection: customer-service skills, attitude, and work product.

Supervisors and managers write a paragraph in support of their nominees. Although it's a small company, the staff members don't know all of the candidates for the award. Prior to the meeting, several managers agree to support each other's candidate in exchange for their vote next time.

Employee recognition programs should not be a popularity contest or based upon how well the person's sponsor embellishes his or her qualifications. There should be specific criteria for choosing the employee-of-the-month. If supervisors or management employees aren't eligible, that should be made clear in the selection criteria.

The selection process shouldn't be a horse-trading session. For the award to be meaningful, it shouldn't rotate from department to department or be given to someone just because that person didn't win before. The recognition process shouldn't depend on whose sponsor writes the best.

Excellence shouldn't go unrecognized. Typically, it's recognized with a plaque or a mention in the company newsletter.

One corporation carries the recognition process further. In a full-page ad in the local newspaper, the company recognizes its Quality Excellence Award Team Members. The employees' names are listed in prominent type.

Recognition doesn't have to be broadcast over the airways or published in the local newspaper. The best managers recognize performance every day with a kind word or a pat on the back.

A survey of human resource executives concluded that praise and recognition help keep employees happy. You didn't need a book or a survey to tell you that. Treating employees with TLC works wonders, especially if it's offered by someone who genuinely cares. Managers should remember birthdays and milestones.

Handwritten thank-you notes go a long way in showing gratitude for work done.

Sharing information with employees and explaining why certain actions are necessary help boost morale. Giving employees credit for suggestions and following through on their ideas are other forms of recognition.

Compensatory time can be useful in recognizing employees who've gone above and beyond the call of duty. Make sure these employees don't lose the benefit of the time off by returning to a huge pile of work on their desks.

In some companies, a prime parking space is reserved for the employee-of-the-month. This can be a great form of recognition, especially when the regular employee parking lot is out on the frozen tundra.

At least one car company gives employers an opportunity to recognize outstanding workers. The employee-of-the-month gets free use of a Ford Taurus or Chevrolet Lumina for 30 days.

Some companies recognize their employees with an evening they won't forget. A bank in Akron, Ohio, recognizes its top performers with a party at the Rock and Roll Hall of Fame in Cleveland. The employees also get shares of company stock.

In some companies, employees are recognized with one-dollar coupons that are good in the company cafeteria. Depending upon the cafeteria, that might not be recognition worth receiving.

As additional food for thought, you might consider rewarding employees and the community at the same time. Employees can be recognized with gift certificates at local restaurants, as well as other businesses.

You can fill up a basket with items from a dollar store or with lottery tickets. The employee to be recognized makes a selection from the basket. When recognition is a part of the culture, you may find yourselves with employees who won't quit if they pick the winning lottery ticket from that basket.

Q-TIPS

Employee recognition is the responsibility of every manager and supervisor on a daily basis. They should foster a culture where employees recognize each other and show gratitude for a co-worker's help. Recognition should be ongoing and consistent.

Every company should have a broad-based recognition program. The recognition program shouldn't, however, be a substitute for standard compensation programs. Employee recognition programs should be a part of a company's total compensation package.

There should be recognition for team achievements, not just individual accomplishments. The recognition program should encourage empowerment and innovation by employees. The employee's efforts to improve quality and customer service should be recognized and rewarded.

CHAPTER 41

Employee Satisfaction

Recognition is a positive act. Some insurance companies, however, send a negative message to a large percentage of their employees. They don't seem to recognize that the people out in the trenches have feelings too.

Whether an insurance company's home office is in Hartford, Connecticut, or Des Moines, Iowa, not every employee works there. It should be a concept that's easy for the folks in the home office to understand: We have employees across the country, if not the world.

Yet frequently, the home office newsletter will focus on the improvements at headquarters, as if the other locations are unimportant. The publication will discuss the new employee fitness center or an all-employee meeting that the CEO will be attending. There'll be cake and coffee and bubkis for the folks in the field.

That's right, bubkis. It's slang for "nothing" or "zilch." Too many companies seem to ignore a large number of employees and make them feel like orphans.

All the great morale-builders and team-building exercises frequently leave some employees feeling left out. The unveiling of the new fitness center could have been announced in conjunction with a program where field employees receive money if it's used toward a health club membership in their area.

Similarly, an insurance company may announce a program where the CEO will have lunch with selected employees. The CEO's down-home approach is announced with much fanfare in the company newspaper. The article never mentions that employees in other states won't be participating in the program.

Care should be exercised so that morale-boosters don't inadvertently offend a large segment of the employee population. A high-ranking insurance executive traveled to the field offices to meet with groups of employees. The employees were picked on the basis of who would be less likely to ask tough questions so management wouldn't be embarrassed.

While you can't please everyone, you should be able to demonstrate fairness in distributing perks of this kind. Management should articulate why particular employees were chosen to attend.

In an atmosphere of Total Quality, perks are distributed throughout the organization. The so-called "soft perks" don't cost anything and are available to everyone, not just managers or a chosen few. Casual-clothes days, for example, have caught on at many companies.

Of course, one claims department manager had his own particular kind of morale problem. He vowed to go on a coffee break with every employee. He was so beloved that most employees prayed their name wouldn't be chosen. A few called in sick on the day of the coffee break with the manager.

It's not just field employees who aren't getting their fair share of the pie. Sometimes, it's certain groups of employees who seem to be excluded.

In the cover story of the business section of *The Miami Herald*, the newspaper lists Florida's highest-paid corporate executives. No woman makes the list of the top 50, which is cut off at $1.2 million in total compensation.

The top female executive among Florida's public companies earned $529,086. Not too shabby, mind you, but nowhere near the top of the list.

In fact, only 3 percent of the 730 highly compensated executives were women. That figure falls right in line with a 1995 Glass Ceiling Commission report which found that 3–5 percent of senior managers in major corporations were women.

At a major insurance company on the list, five executives are listed with salaries ranging from $298,752 to $938,946. All of the executives are men.

In some instances, attitudes are more subtle. An executive always forgets to put the lone female staff member on distribution for documents. Maybe it's just an oversight, a personality conflict, or baggage from his recent divorce. Or maybe it's something more.

Whether it's the insurance industry or any other industry, a culture should exist where women and minorities are respected, as well as nurtured. The glass ceiling shouldn't exist for any employee, regardless of sex, color, religion, or ethnicity.

In the early 1970s, when women were making significant inroads in the work force, few held positions in senior management. "It won't be that way in 20 years," executives would promise. "There will be plenty of women with years of experience under their belts who will be able to move into senior management."

Recently, the Center for Creative Leadership in Greensboro, North Carolina, found that female managers are more effective than men. The study found women to be better managers in 28 of 31 categories. Despite that fact, women managers didn't make the top 50 in Florida.

As companies strive for quality and employee satisfaction, the journey requires a culture where growth for every employee is based upon his or her ability, not a good-old-boy network or some equally reprehensible basis for promotion. Statistics aren't necessarily reflective of the culture that exists in a particular organization, but they are hard to ignore.

Changing the corporate culture is far from easy. Change has to come from the top, and senior management will usually resist it.

In his excellent book *Buyer Beware*, Morey Stettner offers advice on how insurance companies can improve service. Stettner believes that corporations have personalities. He suggests asking employees to describe the organization's character. Do they use words like "caring," "professional," and "efficient," or do they use more negative descriptions?

Stettner believes customers view the corporation in much the same way as employees. He recommends that senior managers focus everyone's attention on the most desired personality traits, and customers will soon notice the difference.

At too many insurance companies, the image they project in advertisements isn't the way they operate. While they try to create a warm and fuzzy image in commercials, these companies play hardball with their customers and with their employees.

Most companies talk of their all-consuming concern for their employees. It's only behind closed doors that management's real feelings come out.

They're all atwitter in the human resources department. An employee tripped and hurt herself during the lunch hour. The human resources department's first concern isn't whether she was hurt or not hurt. The first question is whether the accident is covered by workers' compensation.

They call the lawyers and ask whether the injury happened in the course of employment, since the accident occurred outside during a lunch break. They breathe a sigh of relief when the lawyer says it's not a compensable injury because the injured worker is an hourly employee and the accident happened outside.

You would think that now they are ready to show concern for the injured worker. Well, not quite. Someone brings up the OSHA issue. "Oh, my goodness," they realize, "we may have to fill out an OSHA report."

Now, they're really ticked off at the stupidity of this employee. They may have to fill out an OSHA 200 report, which looks bad for them at headquarters.

While you can use the quality improvement process to make the workplace safer, insensitivity will undermine any quality initiative. The culture must change before any progress will be made.

Employee health and safety should be a top priority for management, and not just to help curb insurance premiums. Wellness and employee assistance programs send a caring message to employees, as well as provide the secondary benefits of reduced absenteeism and lower health insurance costs.

Q-TIPS

Insensitivity to the feelings and emotions of employees will eventually undermine any quality initiative. Any loyalty to the company will gradually erode.

The world-class organization will use employee satisfaction surveys to drive improvements in the human resources area. Quality companies will conduct exit interviews with employees to determine what improvements are needed. Wherever possible, organizations should collect data on training, grievances, safety, and employee turnover as a means to make improvements in the human resources function.

CHAPTER 42

Downsizing and Employee Satisfaction

Companies send subtle messages that some employees are less valued than others. Businesses also send the message that every employee is dispensable.

In October 1996, Aetna announced one of the largest ever job reductions in the healthcare and insurance industries. Newspapers reported that Aetna would cut 13 percent of its work force and would close 30 health insurance claims-processing centers. The reductions followed its merger with U.S. Healthcare.

In his book *The New Doublespeak: Why No One Knows What Anyone's Saying Anymore*, Professor William Lutz says no one gets fired anymore. Employees are "Dehired, deselected, selected out, repositioned, surplussed, rightsized, correct sized, excessed or uninstalled."

Lutz notes that companies don't lay off workers. They make "workforce adjustments," "headcount reductions," "census reductions," or implement a program of "negative employee retention."

Some companies don't even try to put a positive spin on staff reductions. A corporation hires a flamboyant CEO, frequently called the "poster boy for corporate downsizing." The stock of the company soars 53 percent.

The CEO has been called "Rambo in Pinstripes" and "The Shredder," as well as "Chainsaw Al." Reportedly, the CEO is quite fond of those nicknames.

In another era, CEOs would say, "People are our most important asset." There was at least a pretense that they felt bad about downsizing.

The real issue isn't whether downsizing is a necessary evil. The question is how to maintain quality in an environment of downsizing, especially where employees have come to believe that they are dispensable and that their allegiance to the company is probably at its low point.

One problem with downsizing is that it never seems to end at some companies. Instead of making the necessary cutbacks, employees live under the cloud of ongoing reductions in staff.

There's also the ongoing problem of fewer people doing more work. They're already spread too thin and have little time to devote to quality initiatives.

At the Acme Insurance Company, the general liability unit isn't operating at peak efficiency. In fact, it's been this way for months.

The unit manager got fed up four months ago and quit. Another employee retired last month. Bill's out on storm duty and won't be back for two weeks. Two other employees are on vacation this week. Another is out on sick leave.

Who's sorry now? The two employees who are running the unit this week are sorry. That's for sure. A temporary employee is helping out but really won't get the hang of the job for a few days.

Customers are even sorrier. Phone calls aren't returned. No one's working on the files of the employees who are gone. Phones are ringing across the office, but there's no one available to pick them up.

There's no light at the end of the tunnel either. Upper management is dragging its feet on replacing the departed employees. They haven't even interviewed potential candidates yet.

Based upon past experience, the employee who's acting as unit manager won't get the job permanently. She's a 20-year veteran and is unlikely to go elsewhere if she's passed over again for the job.

Employee satisfaction has hit an all-time low in this department. Even if the few remaining employees were motivated to do so, they would find it almost impossible to provide quality service to customers.

At other companies, employees aren't just leaving through attrition. The carrier has cut staff by 10 percent on two occasions this year and won't be replacing any of these employees.

In making these cutbacks, the company ignores its own internal guidelines for file handlers. Instead of handling a suggested case load of 50 new notices per month, employees are now working on over 100 new files each month.

Many employees live in fear of getting downsized. The atmosphere is similar to a recent *Cathy* cartoon. Cathy's co-workers are going on vacation. They leave their phone numbers with Cathy, in case there are any rumors about restructuring, cutbacks, or downsizing. In the last frame of the cartoon, Cathy makes the following observation: "Most popular vacation destination of 1996: Paranoia-Ville."

And as the cuts continue, quality and customer service suffer.

Q-TIPS

The reality of doing more work with fewer people doesn't have to be sugarcoated by management. Nevertheless, employees should be given a voice in how to handle the work more efficiently.

Making false promises to employees only exacerbates the problem of downsizing. Too many employers think it builds morale to have employees believe that reinforcements are on the way and that additional help is forthcoming. Employees perceive that the company is attempting to save money by delaying the hiring of a replacement worker.

To save costs, insurers should be flexible and innovative. One large insurer has a pool of retirees who are willing to come back to work if called upon. There are also employment agencies that specialize in providing temporary insurance personnel.

CHAPTER 43

Bad Employees

Whereas some companies downsize helter-skelter, Insurers R Us never lets anyone go. Now there's a quality company, right? Not!

Wilma's still working for the company, despite years of attendance problems. Her co-workers joke that she's on double secret probation for the eighth time. When she gets in trouble, Wilma cleans up her act for 90 days and is taken off of probation. During that period, she has a miracle cure for her illnesses that besiege her again after she's no longer on probation.

Her excuses are creative, to be sure. She should, however, keep better track of them. Aunt Wilma has died for the third time. This week, she's suffering from sea lice, which is tough for someone in Iowa to contract.

Wilma could have created a sweatshirt sold by novelty shops. The sweatshirt says, "I don't have any sick days left, so I'm calling in dead."

Employees like Wilma cause managers to wake up in the middle of the night in a cold sweat. While downsizing can destroy morale and can have a detrimental effect on quality, hanging on to an undesirable employee is worse. Management does no one a favor by putting up with bad performance.

It's not like Wilma does a bang-up job when she's there. Certainly, her customer-service skills are less than exemplary but wouldn't warrant her dismissal. Her attendance does.

Management sends messages to other employees by tolerating the bad employee. Certainly, there are better ways to motivate than by fear of firing, but being lax sends an even more negative message.

Chronic absenteeism creates a morale problem, especially when co-workers must do their own jobs as well as Wilma's. There is a danger that others may imitate Wilma if they see she gets away with absenteeism. Wilma's attendance problems should be documented and dealt with by management.

If absenteeism is a problem throughout your organization, you can set a goal of improving attendance. You might begin by adding up the total number of employee absences during the past year. It's probably a good idea to exclude extended absences that require a doctor's excuse or time off pursuant to the Family and Medical Leave Act.

With this figure as your baseline, you can measure whether the number of absences has declined. If it has, recognize the group and the accomplishment. You should also recognize individuals who keep their absences to a minimal level.

If you're satisfied with the absence level as a whole, you don't need to focus on that goal next year. You do, however, need to remain focused on Wilma's attendance issue.

Bad managers should not be tolerated either. A manager working on a side business on company time is given a slap on the wrist. A supervisor who is too hands-on with female subordinates isn't disciplined. Another manager makes a racial joke and is permitted to keep his job.

Every company must have clear policies on the type of conduct that won't be tolerated in the workplace. Violations of those policies must be dealt with forcefully and immediately. It's the right thing to do from the quality perspective, as well as from a legal perspective.

Fear of a wrongful-discharge suit is no justification for a company to accept employee misconduct or poor performance. Accepting mediocre performance, or worse, results in poor quality and bad customer service. That's something your customers shouldn't be forced to accept.

Q-TIPS

Attendance is an area of performance that can be easily measured. Trends and improvements will be seen quickly. Absenteeism should be a key element of the performance appraisal. As they say in football, you can't help the team if you're always in the whirlpool.

Discipline must be both fair and consistent. Whereas flexibility is important in an organization, there are certain infractions that should not be overlooked under any circumstances.

Chapter 44

Bad-to-the-Bone Employees and Agents

It takes several years, but they finally catch him. One of the company's highest rated supervisors has been caught skimming money off of claims files.

One day, he's escorted outside the building by an armed guard and someone from the home office. His files are shipped immediately to a group that will audit each and every one of them.

The former supervisor is never prosecuted, even though thousands of dollars are missing. Prosecution, they decide, would be too embarrassing to the company.

The dismissed employee winds up at another insurance company where, hopefully, he's cleaned up his act. The new employer has no knowledge of his past indiscretions.

As part of the hiring process, the new employer did a cursory background check. The human resources manager contacted his references and the former employer. The former employer is only permitted to confirm that the supervisor worked for the company and when he was employed. There is no mention of the illegal activity that led to his dismissal.

Although no one else at that insurance company is skimming money from files, the line between ethical and unethical behavior is blurred. A claims representative socializes on a regular basis with a partner of a law firm she uses and accepts tickets for plays as well as concerts. She's been taken to dinner on a half dozen occasions at expensive restaurants and has even gone boating with the attorney. In return, the claims representative uses that law firm on most of her files.

The problem is certainly not endemic to casualty-property claims. An unethical life insurance agent has been cheating customers for years. Several policyholders complain to the company and to the state insurance department. When the insurance company revokes his authority to write business for it, the agent ties in with another insurer.

The state insurance department suspends his license for a while. During that time frame, he still sells policies and uses other agents to write the business, while collecting most of the commissions.

Previously, we discussed the American Council of Life Insurance's new Principles of Ethical Market Conduct and an accompanying Code of Life Insurance Ethical Market Conduct. Although the Principles and Code only apply to individually sold life and annuity products, you can look at these standards for guidance on what constitutes a thorough background check.

The focus of **Principle 2** is on the distributor. To comply with this principle, the insurer must develop policies and procedures to ensure that its distributors are competent and service oriented.

Here's what the insurer should expect from a distributor:

- Good character and business reputation, and appropriate qualifications and experience.
- Appropriate licenses or other qualifications under state law.
- Adequate training to focus on customers' needs and objectives.
- Adequate knowledge of the insurer's products and their operation.

- Training in the need to comply with applicable insurance laws, regulations, and the Principles and Code of Ethical Market Conduct.
- Participation in continuing education programs.

To comply with the ACLI Principles and Code, the insurer must have documented policies and procedures for investigating distributors. The company must have standards that its distributors should meet.

The assessor will be looking at the appointment standards and the criteria used. The criteria will be much more than having an insurance license and a pulse. Here are a few standards that might be used by an insurance company in its policies and procedures:

- Professional credentials
- Insurance organization membership
- Financial standing
- Any litigation involving the distributor
- Regulatory sanctions, such as a license suspension
- National Association of Insurance Commissioners' producer database
- National Association of Securities Dealers' complaint history

Principle 2 recognizes that the insurance licensing process is insufficient to ensure that a salesperson is competent and customer focused. Many of the requirements of Principle 2 also apply to employees of the insurer who are involved in the sales process and are not just distributors.

The Principles and Code of Ethical Market Conduct must be reflected in the insurer's policies and practices. The company must fully support the concepts in the Principles and Code in a written statement of values. That statement of values must be adopted by the insurer and communicated to employees and distributors.

The insurer must stress the importance of these principles and let people know that noncompliance will not be tolerated. The assessor will look for evidence that ethical market conduct is part of the individual's performance expectations.

Employees and distributors must be told in no uncertain terms that the statement of values is to govern their conduct. Failing to abide by it will have severe consequences, ranging from additional training to dismissal or contract termination.

There is compelling logic for applying the Principles and Code to the hiring process of any insurance company. The aim of Principle 2 is to provide competent and customer-focused sales, and that's a worthy goal of every member of the insurance industry. Obviously, you can't investigate someone who doesn't have a securities license or a license to sell insurance as thoroughly as a professional who does. Nevertheless, insurers do have investigative tools at their disposal that they never use in the hiring process.

Business ethics codes or values statements also serve a useful purpose. Employees should receive mandatory training dealing with the business ethics code or values statement. It might also be worthwhile to set up a hotline to report wrongdoing. Violations of the code or statement of values must be dealt with quickly and forcefully.

Q-TIPS

Every insurer needs to formulate a business ethics code or statement of values. It should be communicated to every employee and distributor. There should be sanctions for violations. There should not be selective enforcement of the code or values statement.

From the first day of their employment, new employees should be made aware of the code of conduct. Management's commitment to the code or statement of values should be demonstrated on an ongoing basis.

CHAPTER 45

Proactive Customer Service

Bad employees and bad-to-the-bone employees are barriers to quality. Whereas it's important to empower employees, some don't deserve to be empowered.

Dr. Richard Hodgetts, a professor at Florida International University, speaks about the 85/15 rule of quality. Eighty-five percent of the mistakes in an organization are the result of ineffective rules, policies, and procedures. Fifteen percent are the result of people who are poorly trained or those who have a bad attitude.

Even as you make progress toward resolving employee issues, you'll still have a large percentage of mistakes to worry about. Quality initiatives can help solve those problems.

Quality problems occur in every industry, and it's useful to examine them. A woman who's had breast cancer and is taking Tamoxiphin notices that she's bleeding. Since uterine cancer is a side effect of the drug, she panics and calls her doctor. Her gynecologist makes her come in for a biopsy. Day after day, the woman waits for results. Despite several calls and assurances that no news is usually good news, the woman can't stand the wait. It's been over a week without a report from the lab.

She visits her gynecologist, who calls the lab. They fax the results to the physician, and fortunately, the biopsy is negative.

More than likely, if the woman hadn't taken the initiative, she would have waited even longer for the results. Although the lab usually sends what's known as a "wet reading" to the doctor if something is wrong, the patient can't relax until hearing the results.

Sometimes delays are compounded by glitches along the communication chain. A slow lab sends results to a physician who is off for a few days or is painfully slow in getting results to patients. The result is added stress for the patient and bad service.

At many insurance companies, there are too many slowdowns along the way. A customer's problem works its way up from the bottom of the pile on one employee's desk to the bottom of the pile on another employee's desk. Somewhere during that journey, supervisors put it on their own pile of things to do, or the mailroom takes its good old sweet time getting the paperwork from one pile to the next.

At too many insurance companies, customer service is reactive, not proactive. The only way to reach the top of the pile is by calling and making a fuss. Employees only deal with pots that are boiling over, not the ones on the back burner.

Calls are routinely initiated by customers, not insurance company employees. File handling is tickled by phone calls, not by a diary system. The end result is ticked-off customers, not ones who are tickled pink.

Return phone calls are slow in coming, if at all. The person the customer needs to speak with is seemingly always on the road, in a meeting, or on the road at a meeting.

Too often, the person taking the message is left making excuses for the lack of communication. While the message-taker is unfailingly polite, the customer is every bit as frustrated.

Even the voice mail messages are frustrating to customers. The message says, "I'm either out of the office or away from my desk, but I'll get back to you as soon as possible." From the message, the caller is never quite sure if the employee went for coffee or to Cameroon for vacation.

There's always the infamous message that gives the name of someone who can be called if there's an emergency. That frequently gets you to the person's voice mail.

Although it's a given that communication is of the utmost importance in any business, you're frequently dealing with emergency situations in the world of insurance. Customers are without a car or a roof over their heads. Someone is injured or needs to go to the hospital. There are dozens of things that need to be done, more pressing things than waiting by the phone for the insurance company to call.

Health insurance policies have managed-care requirements. Insureds call for permission to undergo medical treatment of some kind. It's a stressful enough time without being aggravated by an insurance company's lack of communication.

Improving response time can help any organization achieve a higher level of customer satisfaction. Start by defining your response-time goals. Measure how long it takes now, and define how long it should take.

In 1994, the Zurich-American Group was looking to improve the quality of service provided to personal-lines customers. The insurer first asked customers of its service center what they considered to be important. When customers said they value a quick response, the insurance company began measuring how many rings it takes for a call to be answered.

To other customers, response time is something other than how long it takes for a call to be answered. The commercial customer may want a quote delivered 30 days prior to renewal or a policy that comes within four weeks of the binding date.

To improve response time, look at the steps involved in responding to the customer. Use a flow chart to determine where the bottlenecks are and how to get rid of them. Eliminate steps that don't make sense. Streamline the process wherever possible.

Come up with an improved process, using brainstorming sessions or some other method. Break down the potential causes of delays into groups, such as policies, procedures, and people. Create solutions to the problems. Finally, implement the new process and make sure it's working properly. Make certain there are no newly created bottlenecks that stem from the new process.

With quality and continuous improvement, the job is never quite finished. Products and services can always be improved. Processes can always be improved.

What's the key to process improvements? According to QualTeam, Inc., it's standardization, simplification, feedback, bureaucracy elimination, and cycle-time reduction.

Process improvements aren't rocket science, just common sense. You're simplifying the process, while keeping your eye on the customer.

You need to decide if every step in the process is worth keeping. Does it add value, or does it just delay the resolution of the situation?

Look at it as job security. There will always be work that needs to be done.

Q-TIPS

Responsiveness is an important dimension of service quality. Employees must be ready, willing, and able to provide service. The service must be performed in a timely fashion.

Keep customers informed as to the status of claims and issues that are outstanding. Communicate your progress to resolve the issue, not just the resolution. Insofar as claims are concerned, the unfair insurance practices act in your state is just a starting point and doesn't represent the ultimate in customer service.

Initiate calls to customers instead of waiting for them to nag you. Phone tag is particularly frustrating. Avoid it by giving employees beepers or some other mechanism for keeping in touch with customers.

CHAPTER 46

Quality Meetings on Quality

Meetings to improve quality and customer service sometimes are the antithesis of quality. They alienate internal customers, the employees.

It's your worst nightmare. They're baacck! It's the invasion of the pinheads from the home office.

It seems as if their full-time job is taking you away from yours. They show up with new directives and programs that must now be implemented. They announce that they're coming and when the meetings are.

And the meetings go on for hours. Rambling aimlessly, they stretch an hour's worth of information into a six-hour meeting as if to justify their trip.

From the perspective of many employees, the quality gurus from the home office live in an ivory tower. They've never been in the trenches or out in real world, where theory doesn't coincide with reality.

When you talk about being at the wrong place at the wrong time, most people think about meetings. If you're overworked, there's never a good time or place for a meeting. And to many cynical employees, meetings to improve quality are the worst.

The situation is aggravated by a meeting that's meaningless. Everyone's talking at once. There's no agenda. No one prepared minutes from the last meeting.

Instead of a productive meeting, you're sitting through another beef session. It's just an ongoing dialogue about what's wrong with the company or the product, not how to fix it.

The person leading the meeting should minimize the training exercises that are designed to spark creativity. There is no time for exercises on how to think outside the box. You're wasting valuable time.

Meetings to improve quality don't have to be that way. While a certain amount of social activity and griping is to be expected, a meeting shouldn't degenerate into chaos.

As workers continue to get more on their plates, meetings should get shorter, not longer. Co-workers can catch up on personal issues after the meeting, not during it. A quality meeting doesn't have to be long and involved.

Q-TIPS

Everyone should leave the meeting with action items. The parties should know what they're responsible for, so they can be prepared for the next meeting.

Minutes should be kept, so that everyone is kept on track. Minutes help focus the participants on the issues at hand.

Meetings should have closure. Issues should be resolved, or there should be steps outlined so that closure can take place at some point. Agreeing to disagree is not closure.

Every meeting should have ground rules. These rules should clearly state the purpose of the meeting. There should be no deviation from those ground rules.

Stick to an agenda. The meeting should start on time and end on time.

CHAPTER 47

Quality Surveys

You can count on it every night at about dinnertime. A telephone solicitor will call saying she's taking a survey. When the first few questions ask about the condition of your rugs, it becomes fairly clear that someone's about to sell you carpet cleaning services and isn't really taking a survey.

The real surveys frequently look like junk mail and are treated as such. In fact, the only time survey forms are in demand is when the hotel has just lost your reservation or you've gotten terrible service in a restaurant and are looking for a way to vent your anger.

Businesses try every trick in the book to take surveys. They'll offer gifts to those who complete the survey or turn the process into a contest. They'll put the names of all the people who respond in a fishbowl, and one of them will win a prize.

Some surveys beat around the bush when asking questions. The hotel chain's survey asks how many nights in a hotel the customer is spending during the year. The chain already knows the customer hasn't stayed there all year. They might as well ask, "Have you stopped traveling, or do you just not like us anymore?"

As a rule, surveys attract the unhappiest of customers. The pleased-as-punch customers just won't find the time or energy to respond.

Designing the perfect survey is extremely difficult. Too many questions will discourage people from responding. Too few questions result in insufficient data. Questions must be well focused to elicit the information you're seeking.

After making a change to a life insurance policy, a USAA Life member received a survey. The customer asked why a survey was sent and what would be done with the information. USAA responded as follows:

> *We conduct surveys for several reasons. First, we use them to gauge what products and services our members want. Many of our new products, as well as changes to our existing products, are the result of information obtained from surveys. Secondly, surveys tell us if our members are satisfied with the service they receive, and if not, how we can improve. Instead of waiting for complaints, we use surveys to gather input so we can avoid potential problems. We send a mail-back survey to a random sample of everyone who conducts a transaction with a USAA Life representative. We know your time is valuable, and we assure you that it is worth your effort to fill out and return these surveys. Every survey we receive is read and recorded by our quality assurance department. We welcome constructive criticism and have used it to make important business decisions. In addition to mail-back surveys, we gather feedback from both letters and telephone conversations. When a member makes a suggestion while talking with a representative, we document these comments so they can be analyzed and acted upon when feasible. Occasionally, we participate in industry studies to benchmark our products and services.*

Based upon the critical acclaim that USAA receives for its policies and service, it seems clear that the company is making excellent use of surveys.

Q-TIPS

Surveys can be enormously helpful in achieving Total Quality, if they're designed properly. Nevertheless, don't forget about the informal surveys. Management should pick up the phone from time to time and call customers at random. A survey doesn't always have to be structured or formal. Personal interviews with customers are extremely useful. Another option is a focus group. A moderator meets with eight to ten customers for an hour or two to find out their opinions about service. Their responses can be used to create questions for a larger survey of customers. The results of surveys, good and bad, should be shared with members of the staff. An effort should be made to find common themes among the comments made in surveys. If possible, management should attempt to review the underlying transaction that led to a negative survey and determine what went wrong.

CHAPTER 48

Quality Relationships

A while back, the musician Prince changed his name to a symbol that no one understood. Radio stations began referring to him as "the artist formerly known as Prince."

Some insurance companies seem to be taking the same approach. They've been bought and sold so many times, customers feel like they're dealing with the insurance company formerly known as Prince.

Even the big insurance companies are making sweeping changes that are confusing to consumers. There are joint ventures between companies like The Travelers and Aetna. HMOs buy each other out and merge.

Consumers find an old policy when a relative dies. They attempt to trace who is responsible for the death benefit. After calling around for days, they find out that the original company was bought out, as have three previous insurers who once had their hands on the policy.

Quality and customer service have to remain constant, even as the business climate changes. When banks merge, customers usually come in the same office, but there's a different name across the front door. Frequently, they're greeted by the same personnel.

When insurance companies merge, there may only be a single letter that goes out to customers. The letter may get filed away or not filed at all, since it's often sent bulk rate and will look like junk mail. Even the content of the letter suggests that the change is meaningless. The new insurance company downplays that the old one has disappeared off the face of the earth.

Business conditions also result in insurers that used to sell particular lines of insurance. An insurance company will no longer do business in a particular state because economic conditions warrant withdrawal. Long-time customers are left to scramble for coverage because their insurance company no longer writes auto or homeowners insurance.

Rebuilding that relationship again is almost impossible. Customers will always believe that you'll leave them in the lurch again if it's in the insurance company's best interests.

It's very much like the problems baseball faced after the strike. Many fans lost their desire to commit time and money to following the sport. Baseball was forced to develop a new marketing strategy and brought in a former Kraft Foods executive for guidance. Previously, the executive helped Kraft restore the Kool-Aid brand, which was hurt significantly by the Jim Jones cult suicide in Guyana.

Many insurance companies seem to be committing customer-service suicide by withdrawing from a state due to adverse business conditions. The decision has an extremely negative impact on customers who may have other policies with the company. In a business built on trust and relationships, withdrawing from the state sends a terrible message to customers of that company.

The carrier that values customer service may come up with a compromise position. That position might be to honor the contracts it has now and renew those policies, but the company won't write any new business in the state. Carriers, on the other hand, will argue that customer service doesn't mean sending the insurance company into insolvency.

Recently, Nationwide Insurance Companies began trimming sales of new policies in the coastal areas of 16 states, as well as the District of Columbia. The decision to stop selling new policies came on the heels of substantial losses from Hurricane Fran.

As insurance companies leave certain lines of insurance behind, as well as loyal insureds in certain states, they walk out on good customers who will never want to deal with them again. These are not the only ways to antagonize customers. With apologies to Paul Simon, who said there were 50 ways to leave your lover, there are at least 50 ways to lose your customers. Here are a few:

Don't return calls, Paul.

The check's in the mail, Gail.

Screw up the bill, Will.

Butcher their last name, Wayne.

There are many more ways to alienate customers, but we'll leave it up to you to rhyme them. Here are some of them:

Don't meet deadlines.

Make excuses.

Renege on commitments.

Keep them on hold.

Instead of singing this tune, you should be looking for better advice on customer service from Paul Simon. Look on the *Bridge Over Troubled Water* album for the Simon and Garfunkle song called "Keep the Customer Satisfied."

Not many people realize Mick Jagger was talking about customer service when he pranced his way through "Satisfaction." Okay, so it has nothing to do with quality, but don't be a spoilsport.

The Righteous Brothers were talking about how difficult it is to court back former customers in "You've Lost That Loving Feeling." The Temptations tried to woo them back, to no avail, as they lamented in "Ain't Too Proud to Beg."

You've probably noticed by now that there are no songs from the 1980s and 1990s, since we haven't heard a song we liked in the past two decades. All kidding aside, many insurance companies manage to end relationships with customers on a sour note, no pun intended.

Q-TIPS

Insurers have ongoing obligations to the customers of companies they take over. There should be annual correspondence with those customers, reminding them that they now service policyholders of the previous insurance company. Not only is this the right thing to do from a customer-service standpoint, it makes good business sense since these are prospects for new policies and products.

Instead of dropping all policyholders in a particular area, the better approach is to stop writing new business. This practice imposes less of a hardship on existing customers. If a company is withdrawing from a state, it should seek out another carrier to assume the policies of existing clientele.

CHAPTER 49

Keeping Internal Customers Satisfied

Keeping internal customers satisfied is every bit as important as satisfying the external ones. Bob thinks he's finished for the week. There's a laugh. Little does he know that the marketing department has a presentation on Monday and will be dropping by at any moment to ruin his weekend.

The marketing department has known about the presentation for weeks. Nevertheless, Bob will be hearing about it for the first time in the next few minutes. The presentation requires numbers only Bob can provide.

It's not the first time that Bob's weekend has been ruined by another department. Bob plans his day and week accordingly and suffers when someone else doesn't do their job right.

Worse yet, the marketing department has a credibility problem. There have been times when they gave Bob an artificial deadline, only to admit later that they really didn't need it so quickly.

On several occasions, Bob broke his back to finish the rush project and was told later that the presentation was postponed. He went to his boss to complain about the ongoing situation and got little satisfaction. After all, Bob's boss wants to be viewed as a team player so he can stay in the good graces of the marketing manager.

Sometimes poor planners wreak havoc on their own departments. A boss or supervisor waits until the last minute to assign work, and the rest of the group suffers. On other occasions, they can drag down several departments with them.

There should be solutions other than entering the supervisor in the world's worst boss contest that one company sponsors every year. Aside from the employee satisfaction issue, there's the ongoing quality problems that result from rushing work out.

In many organizations, there is infighting between departments. Their mission is the same, but they can't agree on how to achieve it. Neither wants to believe that the other is a customer and should be treated as such. No one can seem to convince them they're on the same team.

Teamwork problems won't be solved with a few team-building sessions. After the glow of team-building sessions dies, the romance turns to cold ashes.

A publication called *Successful Women* offered a great explanation of teamwork in a short piece that didn't mention the author's name. It starts like this:

> *When geese fly in formation, they travel 70 percent faster than when they fly alone. Geese share leadership. When the lead goose tires, she goes back into the "V" and another flies forward to become the leader. Geese keep company with the fallen. When a sick or weak goose drops out of flight formation, at least one other goose joins to help and protect.*

The piece goes on to mention that we can accomplish much more, much faster by being part of a team.

It's too bad quality improvement teams can't follow the lead of geese. The members peck at one another as if they each have their own agenda, not the team's. They squabble over trivial issues and bicker among themselves.

When employees of an insurance company work as a team, the insurer does the little things that other organizations don't. When a member calls USAA to change an address, the representative passes it along to all of the company's divisions. There's no need to call every division that handles a member's policy, whether it's life, health, auto, or homeowners. If the member has purchased investment products through USAA, that division will also be notified.

The organization that operates as a team is the best at providing service to the individual customer.

Q-TIPS

Team players don't use the "not my job" excuse. They don't pass the ball and walk off before seeing if it's caught.

Teamwork springs from feeling as if you're a stakeholder in the company. Team players take ownership of calls and attempt to put them in the hands of people who can solve the problem.

The corporate culture should encourage teamwork. Performance reviews should reflect the individual's performance as a team member. Employee training programs should reinforce the concept of teamwork. The culture should recognize and reward the members of a team.

Chapter 50

Quality Versus the Competition

In the movie *Butch Cassidy and the Sundance Kid,* Paul Newman and Robert Redford are on the run from a superposse. Despite every evasive technique, the posse remains hot on their trail.

Over rocks and boulders, through canyons and mountains, the two outlaws attempt to avoid the posse. As the exhausted men seek to avoid capture, they speculate as to who's after them. They wonder if it's Joe Lefors, the best lawman in the West.

Butch cries out in exasperation, "Damn them anyway. Aren't they hungry? Aren't they tired?"

But the posse is relentless and won't give up the chase. They are tenacious in their pursuit, following the trail no matter where it takes them.

Sometimes dealing with your competitors makes you feel like you're in the same boat with Butch and Sundance. They're doggedly attempting to snatch a bigger market share, most likely out of your market share.

In a competitive marketplace, you need to keep a constant eye on the competition and learn from them. Where do they excel, and how can we gain a competitive edge over them?

It's all part of benchmarking. Butch and Sundance didn't quite have the hang of it.

There's a saying that the worst day at the beach is better than the best day at work. Conversely, on your worst day at work, you should still be better than the competition on their best day.

If you keep shooting to raise the level of quality, compliance, ethics, and customer service at your company, you'll wind up far better off than Butch and Sundance. No posse of angry regulators and policyholders will be on your tail. Instead, a large herd of customers will be breaking a trail to your company.

INDEX

D

E

F

N

O

P

Q

R

S